D08820752

THE MORAL COMPLEXITIES OF EATING MEAT

THE MORAL COMPLEXITIES OF EATING MEAT

Edited by Ben Bramble *and* Bob Fischer

WITHDRAWN

OXFORD
UNIVERSITY PRESS

OXFORD
UNIVERSITY PRESS

Oxford University Press is a department of the University of Oxford. It furthers the University's objective of excellence in research, scholarship, and education by publishing worldwide. Oxford is a registered trade mark of Oxford University Press in the UK and in certain other countries

Published in the United States of America by Oxford University Press
198 Madison Avenue, New York, NY 10016, United States of America

© Oxford University Press 2016

All rights reserved. No part of this publication may be reproduced, stored in a retrieval system, or transmitted, in any form or by any means, without the prior permission in writing of Oxford University Press, or as expressly permitted by law, by license, or under terms agreed with the appropriate reproduction rights organization. Inquiries concerning reproduction outside the scope of the above should be sent to the Rights Department, Oxford University Press, at the address above.

You must not circulate this work in any other form
and you must impose this same condition on any acquirer

Library of Congress Cataloging-in-Publication Data
The moral complexities of eating meat / edited by Ben Bramble and Bob Fischer.
    p.   cm.
Includes index.
ISBN 978–0–19–935390–3 (cloth : alk. paper) 1. Meat—Moral and ethical aspects.
2. Meat industry and trade—Moral and ethical aspects. I. Bramble, Ben, editor.
TX373.M67 2015
338.1'76—dc23
                        2015004578

9 8 7 6 5 4 3 2 1

Printed in the United States of America on acid-free paper

CONTENTS

CONTRIBUTORS

Chris Belshaw teaches philosophy at the Open University and also at the University of York. Most of his work is at the theoretical end of applied ethics, concerning the value and meaning of life, the badness of death, and questions about the self, the environment, and the future. He lives mostly in England's Lake District, alongside sheep, cows, deer, rabbits, pigeons, buzzards, and so on.

Ben Bramble received his PhD in philosophy from the University of Sydney in 2014. He is a Postdoctoral Fellow in Practical Philosophy at Lund University. His main research interests are moral and political philosophy.

Donald Bruckner is Associate Professor of Philosophy at Penn State University, New Kensington. His research pursuits are mainly in human well-being and practical reason; he is also interested in animals, food, and the environment. His work appears in such journals as *Australasian Journal of Philosophy*, *Philosophical Studies*, *Utilitas*, and *Journal of Social Philosophy*. He collects road-killed deer for personal consumption near his small farm in rural Pennsylvania.

Mark Budolfson often works on interdisciplinary issues at the interface of ethics and public policy, especially in connection with dilemmas that arise in connection with common resources and public goods.

J. Baird Callicott is University Distinguished Research Professor and Regents Professor of Philosophy at the University of North Texas. He is the co-Editor-in-Chief of the *Encyclopedia of Environmental Ethics and Philosophy* and author or editor of a score of books and author of dozens of journal articles, encyclopedia articles, and book chapters in environmental philosophy and ethics. Callicott has served the International Society for Environmental Ethics as President and Yale University as Bioethicist-in-Residence. His research goes forward simultaneously on four main fronts: theoretical environmental ethics, comparative environmental ethics and philosophy, the philosophy of ecology and conservation biology, and climate ethics. He taught the world's first course in environmental ethics in 1971 at the University of Wisconsin–Stevens Point. He is currently Visiting Senior Research Scientist in the National Socio-environmental Synthesis Center, funded by NSF.

Julia Driver is Professor of Philosophy at Washington University in St. Louis. She specializes in normative ethical theory, moral psychology, and Humean accounts of moral agency. She is the author of *Uneasy Virtue* (Cambridge, 2001); *Ethics: The Fundamentals* (Blackwell, 2006); and *Consequentialism* (Routledge, 2012). She has published articles in *Journal of Philosophy, Australasian Journal of Philosophy, Philosophy, Ethics, Noûs, Philosophy & Phenomenological Research*, among other journals. She has received a Laurance S. Rockefeller Fellowship from Princeton University, a Young Scholar's Award from Cornell University's Program on Ethics and Public Life, an NEH Fellowship, and an HLA Hart Fellowship from Oxford University.

Bob Fischer is Assistant Professor of Philosophy at Texas State University. He works on issues in applied ethics, moral psychology, and modal epistemology.

Lori Gruen is currently Professor of Philosophy as well as of Feminist, Gender, and Sexuality Studies and of Environmental Studies at Wesleyan University, where she also coordinates Wesleyan Animal Studies and chairs the faculty committee for the Center for Prison Education. She is a Fellow of the prestigious Hastings Center for Bioethics. Professor Gruen has published extensively on topics in animal ethics, ecofeminism, and practical ethics more broadly. She is the author of three books on animal ethics, including *Ethics and Animals: An Introduction* (Cambridge, 2011) and *Entangled Empathy: An Alternative Ethic for Our Relationships with Animals* (Lantern, 2015). She is the editor of five books, including *Ecofeminism: Feminist Intersections with Other Animals and the Earth* with Carol J. Adams (Bloomsbury, July 2014) and the *Ethics of Captivity* (Oxford, May 2014) and the author of dozens of articles and book chapters.

Robert C. Jones is currently Assistant Professor of Philosophy at California State University, Chico. He is also a member of the Advisory Council of the National Museum of Animals and Society and a speaker with the Northern California Animal Advocacy Coalition. Professor Jones has published numerous articles and book chapters on animal ethics, animal cognition, and research ethics, and has given nearly forty talks on animal ethics at universities and conferences across the globe. He was a post-doctoral fellow at Stanford University, a visiting researcher for the Ethics in Society Project at Wesleyan University in Connecticut, and a Summer Fellow with the Animals & Society Institute. He lives in Chico, California, a small ag community in Northern California, where he spends time arguing animal rights with local cattle ranchers.

Neil Levy is Professor of Philosophy at Macquarie University, Sydney, and Research Fellow at the Uehiro Centre for Practical Ethics, University of Oxford.

Clayton Littlejohn is a Lecturer in Philosophy at King's College London. He works on issues at the interface of ethics and epistemology and is currently finishing a book on metaepistemology.

Tristram McPherson works in ethics and its semantic, metaphysical, and methodological foundations. He has work published or forthcoming in venues including *Journal of Ethics and Social Philosophy*, *Journal of Moral Philosophy*, *Mind*, *Noûs*, *Oxford Studies in Metaethics*, *Pacific Philosophical Quarterly*, *Philosopher's Imprint*, *Philosophical Books*, and *Philosophical Studies*. He teaches at Virginia Tech. Before coming to Virginia Tech, he received a BA from Simon Fraser University and a PhD from Princeton University and taught at the University of Minnesota Duluth.

Alexandra Plakias received her PhD from the University of Michigan and is Assistant Professor of Philosophy at Hamilton College.

INTRODUCTION

Ben Bramble
Bob Fischer

Peter Singer published *Animal Liberation* in 1975; since then, the ethics of eating meat has been a prominent topic in moral philosophy. It's uncontroversial that the animals we eat are sentient beings; there is "something that it is like" to be them; they can sense the world around them; they can feel pleasures and pains. What's more, they can want things in their environments, form memories, solve problems (sometimes quite sophisticated ones), experience various emotions, and empathize with others of their kind. Nevertheless, each year we bring billions of these sensing, feeling, thinking beings into the world so that we can use their bodies for our own purposes: we cook and eat them; we wear them; we perform tests on them; we make countless products from what's left over. And most of the time, the inner lives of these beings don't register on our collective radar. They are out of sight and out of mind, the nameless backstories to the goods we enjoy. Of course, we don't want these animals to suffer unnecessarily. But judging by standard US farming practices, this desire doesn't run very deep.

The above isn't exactly an argument against industrial animal agriculture—or animal agriculture of any other kind. We've just made some observations. When we lay out the details, will they establish that these practices are morally wrong? And depending on our answer, how should we respond?

The aim of this collection is to explore these two questions. To that end, we've gathered twelve new essays. Some are by ethicists who are theoretically oriented, while others are by those who focus on applied questions; some of them are by philosophers who have written a great deal about these topics, while others bring fresh perspectives to the debates. Our hope is that these essays will advance a pressing moral conversation.

To be clear, none of our contributors defend the status quo in US agriculture. Rather, they take up three issues. First, some wonder how much the arguments against meat-eating establish. Let's grant that, when we consider the meat actually available to us, we should abstain. But perhaps there isn't anything *intrinsically* wrong with eating meat; perhaps the problem isn't with using animals, but with the way we treat them. So we might investigate whether any form of sustained animal agriculture is morally permissible, or whether there are non-agricultural means by which we could eat meat morally. Second, though we might be convinced that the arguments against meat consumption are on the right track, we might doubt that they're correct in their details. Consider, for example, the thin connection between the actions of an individual consumer and the suffering of any particular animal. How does our apparent causal impotence affect our moral culpability? And third, even if we grant that the arguments succeed, we might raise concerns about how we should navigate a carninormative world. How should we understand our identity as abstainers? And how should we relate to those who don't share our values?

Correspondingly, the book has three parts. The first part features essays by philosophers who are running against the grain. These philosophers argue that meat-eating is sometimes morally permissible, and perhaps even in certain cases morally required.

The second part features essays that attempt to improve or build upon existing arguments for vegetarianism, or respond to some of the major arguments against vegetarianism.

The final part of the book features essays that consider the significance of the debate over eating meat. Suppose, for example, that the arguments succeed. What would this mean for the sort of people we ought to become? Alternately, suppose the debate is intractable. Might it be valuable anyway? If so, how?

In the rest of this introduction, we provide a brief overview of the contributions. Part I begins with Christopher Belshaw's "Meat." Belshaw doesn't claim that we benefit animals by bringing them into existence, or that this may form the basis of a duty to create them. But he does argue that it is permissible to bring certain kinds of animals into existence and then permissible again to kill them. It is permissible to kill them because, while such animals can lead good lives, they cannot want to live these lives. Once these animals are dead, Belshaw claims, it is permissible to eat them.

In "Strict Vegetarianism Is Immoral," Donald Bruckner tries to turn a standard argument for vegetarianism on its head. Suppose it's wrong (knowingly) to cause, or support practices that cause, extensive, unnecessary harm to animals. You might think that, given as much, it's always wrong to eat meat. But as Bruckner points out, some plant consumption causes extensive harm to animals, and it's unnecessary insofar as we have other dietary options. And some of us *do* have other options: we could eat roadkill. So, he argues, we ought to eat it.

J. Baird Callicott's "The Environmental Omnivore's Dilemma" offers an alternative to the standard consequentialist and deontological approaches to animal ethics. Callicott advocates a kind of moral pluralism, with overlapping spheres of moral concern, each of which is organized by different principles and obligations. So in addition to our local, parochial concerns, we are also members of a national community, as well as a larger human community, and ultimately the biotic community as a whole. On his view, we don't owe exclusive allegiance to any one of these communities, and we have to weigh their demands against one another. Thus, there may well be circumstances in which local relationships make meat-eating permissible (if not required), even though industrial agriculture is problematic.

In Part II, we turn to essays that attempt to improve or build upon existing arguments for vegetarianism, or respond to some of the major arguments against vegetarianism. We begin with three articles focused on the so-called "inefficacy" objection, the worry that we aren't obliged to become vegetarians since "an individual's decision to consume animal products cannot really be expected to have any effect on the number of animals that suffer or the extent of that suffering, given the actual nature of the supply chain that stands in between individual consumption decisions and production decisions" (as one of our authors, Mark Budolfson, nicely puts it).

In "Individual Consumption and Moral Complicity," Julia Driver argues that even though there are circumstances in which one person deciding to eat meat on a given occasion makes no difference to production policies, such an individual can nonetheless participate and so be wrongfully complicit in the harms of meat production by choosing to eat it. One can be blameworthy with respect to a bad outcome even if one did not cause it. Moreover, this position, Driver argues, is perfectly consistent with consequentialism.

Mark Budolfson begins his chapter "Is it Wrong to Eat Meat from Factory Farms? If So, Why?" by painstakingly laying out the evidence for the empirical claim at the heart of the inefficacy objection. He then considers an attempt to respond to the inefficacy objection by appealing to the sort of considerations that explain why we have a reason to vote even though our vote is extraordinarily unlikely to make a difference to the electoral outcome. Finally, he proposes his own solution to the problem, which invokes the notion of *the degree of essentiality of harm to an act.*

In "Potency and Permissibility," Clayton Littlejohn considers whether there's anything that can be said in favor of being an "unreflective carnivore." To that end, he runs a version of the argument from marginal cases, according to which animals deserve whatever moral consideration due to "marginal" humans—infants, the severely mentally disabled, those in comas, and so forth. Even if this argument works, though, it doesn't show that we shouldn't eat meat: to reach that conclusion, we need further premises. Littlejohn considers three ways that someone might try to prevent such an argument from getting off the ground. First, he examines the suggestion that while *some* animals meet the bar of moral considerability, the ones we eat don't. Second, he considers the possibility that there may be circumstances in which an individual's action makes no difference to whether an animal dies—for example, in a restaurant in which lobster is served, and any lobster that isn't killed today will just be killed tomorrow. Third, he discusses the same causal-supply-chain worries to which Driver and Budolfson attend, defending a version of the "trigger" solution that they set aside.

In "A Moorean Defense of the Omnivore," Tristram McPherson shifts the conversation. When confronted with arguments against meat-eating, many people find the conclusion incredible: it seems *obvious* to them that meat-eating is morally permissible. McPherson considers one charitable interpretation of this response: namely, that you might think you have better evidence for omnivory's permissibility than for all the premises in an argument against it, just as you might think you have better evidence for the claim, "I have hands," than for all the premises in a skeptical argument. This intriguing move raises difficult questions about how we assess Moorean arguments. To this end, McPherson proposes five criteria that, jointly, we can use to evaluate their strength. As McPherson shows, the Moorean case against the skeptic does fairly well on these criteria. He argues, however, that the case for omnivory fares rather poorly.

Ben Bramble's "The Case against Meat" rounds out this section. In his chapter, Bramble attempts to shore up an intuitive argument against the human practice of raising and killing animals for food (i.e., the argument according to which this practice is extremely harmful to animals, but only trivially good for us). He does so by providing new responses to the four main objections to it: (1) that this practice is *not* extremely harmful to animals, (2) that this practice is far *more* than trivially good for us, (3) that animal welfare is less *important* than human welfare, and (4) the inefficacy objection. He discusses the non-identity problem and the pleasures of meat. Central to his responses to (2) and (4) is the idea that our involvement in meat-eating takes a large psychological toll on us.

"Veganism as an Aspiration" opens Part III. Lori Gruen and Robert Jones note that it is problematic to identify as a vegan, since that self-understanding is often taken to imply that you have clean hands, that your actions don't contribute to animal suffering or exploitation. As they argue, this is mistaken: so many of our actions harm animals, and it is virtually impossible to opt out of them all. If we understand veganism as a kind of moral success, we are setting ourselves up for failure. None of this shows, however, that we can't pursue vegan ideals. They thus propose that we should understand veganism as a kind of aspiration; and so construed, they think that it can make a difference for the animals it's designed to serve. In particular, they argue that this type of veganism can serve as a form of resistance to violence, an example to others, and the foundation for political change.

Neil Levy contributes "Vegetarianism: Toward Ideological Impurity." In it, he takes a page from Jonathan Haidt's work on how rules become sacred. According to Haidt, when a society marshals disgust to enforce a rule, people tend to lose sight of what makes the rule valuable to the society in the first place. Moreover, when people are disgusted by violations, you become defiled if you are the violator. This threatens your standing in the community. Levy maintains, however, that this is unproductive for vegetarians who are concerned about animal welfare. On the one hand, it can turn an isolated lapse into a reason to abandon your commitment entirely. On the other—and more importantly—it can make you less able to support others in their faltering attempts to lower their meat-consumption. On Levy's view, then, we should avoid sacralizing our rules. They ought to be strict, so as to more effectively guide action; at the same time, though, they should be held in such a way that we forgive lapses, thereby making us more effective advocates for animals.

In "Against Blaming the Blameworthy," Bob Fischer assumes that it's wrong to eat meat, and also that most people are blameworthy for doing so. However, he's not convinced that we should blame them: on his view, you shouldn't blame someone for his behavior if it would be unreasonable to demand that he behave otherwise; and in our current context, Fischer argues, it would be unreasonable to demand of most people that they abstain from consuming meat. Fischer's argument is based on the idea that the arguments against meat-eating generalize—that is, if they work, then they show that we ought to live much differently than we do. But since it would be unreasonable to demand that people live up to all these obligations, and arbitrary to insist on any one change without explanation, it's unreasonable to demand that people abstain from eating meat.

We close the book with Alexandra Plakias's "Beetles, Bicycles, and Breath Mints: How 'Omni' Should Omnivores Be?" This chapter is a reflection on the moral significance of the debate about meat-eating, which Plakias takes to be an intractable one. Nevertheless, she argues, it's a valuable debate in which to engage: in so doing, we are also working through questions about the nature of food. As she points out, we can think of different camps in the animal ethics literature as endorsing different theories about what food is: some characterize it in terms of the kind of thing it is (animals aren't food because they're sentient beings), and others characterize it in terms of the process by which it's produced (animals aren't food because they're abused before bringing them to the table). Her hope is that food ethicists can draw on these debates to help us better navigate a world with an ever-increasing number of "food products"—with an emphasis on the second term.

We—the editors of this volume—are vegetarians. This book began with conversations about what could be said against our shared commitment not to eat meat. Those discussions have evolved into the project before you—a mixture of provocation, precisification, and reimagining—all based on the conviction that it matters how we relate to animals. We've gathered these essays together in hopes that they will advance the discussion about what we ought to eat. If they manage to further those exchanges, prompting new conversations and refining existing ones, then they will have served their purpose.

DEFENDING MEAT

1 MEAT

Christopher Belshaw

Introduction

Are we permitted to bring things into existence in order to kill and eat them? Certainly we are. Otherwise we wouldn't, with a clear conscience, eat potatoes. Are we permitted to do the same with animals—start their lives, kill them, and eat them? This is less straightforward. And the major reservations relate not to meat eating[1] as such but to its usual consequence that animals providing the meat thereby live bad lives, or at least worse lives than they would otherwise live. And these lives are bad, or worse, in at least one of two respects. Either they contain more pain than otherwise, or they involve the animal in a premature death. It is allegedly bad for animals to suffer this pain and allegedly bad too for them to die.

---

1. I mean by meat the fleshy parts of animals, including so-called red and white meat, birds and poultry, fish, shellfish, and so forth. I am not here counting the fleshy parts of nuts, fruit, and vegetables as meat, and am not counting those who use animal bones for gelatine as meat eaters, even though they are in fact not vegetarians. I make this point because it is not uncommon to find that fish eaters are classed among vegetarians. And after making what appears to be, and is said to be, an argument for vegetarianism, David DeGrazia (2009) p. 164 insists that he has "no position" on eating fish and invertebrate seafood. To my mind this is tantamount to having no position on vegetarianism.

I'll say more about death than pain. I'll argue, although not at length, that killing animals is in some circumstances permitted, and in some circumstances required. These claims are, I hope, more or less uncontroversial—controversy starts in detailing the circumstances. And I'll argue—more controversially and at greater length—that killing animals is very often permitted and, further, at least suggest it is very often required. Having killed them, we may as well eat them. So then meat eating is permitted.[2] Is it also required? I won't claim this. What I will claim, however, is that it is perhaps required that there be in place procedures and practices that have the production and consumption of meat as one of their primary aims. Or at least, that there are good reasons for sustaining such practices. My concerns here are with morality rather than expedience. And the predominant, but not the only, concern is not with what is good or bad for us, or for the universe, but what is good or bad for animals, and particularly for those we eat.

Meat, Pain, and Death

Someone says, I do nothing wrong in eating meat. The animal is dead, was dead when I cooked it, even when I bought it. It's past caring. This is, of course, mere sophistry. Eating, and wanting to eat, meat has many consequences and plays a pretty direct causal role in several of the practices determining the contours of animal lives. How do our proclivities and their fortunes interact? Consider the various ways we might come by meat:

Synthesizing

Perhaps it is, or soon will be, possible to fabricate meat in a laboratory. Producing and eating this meat will have no direct consequences for any animal. There will be indirect consequences, however. Insofar as such a practice takes off there may well be fewer animals living bad lives, and fewer living any kind of life at all.[3]

---

2. Or some meat eating is permitted. Even if killing human animals becomes in some circumstances acceptable, eating them is likely still to be proscribed. And this could be the case for some non-human animals also.

3. But won't this synthetic meat encourage a taste for the real thing? Some version of this slippery-slope counter-argument can be made to all the methods of meat production sketched here. Such arguments are, I think, overrated. I know, for example, many fish eaters who are no more tempted to eat poultry or red meat than vegetarians.

Sampling

Suppose it is possible to cut and eat meat from a living animal, which then recovers. Suppose that even when repeated, this practice doesn't hasten death. Still, insofar as it causes the animal some pain (as is likely), such a practice is morally dubious, to say the least.

Scavenging

We might eat animals whose pains and deaths are independent of meat consumption and are caused by traffic, other animals, weather, bad luck. Some think that eating such victims of circumstance is altogether morally innocuous. But there are consequences for other animals. Some are thereby deprived of food. Others may lose the opportunity to grieve or to mourn.

Hunting

Perhaps there are some instances of a clean kill, but typically hunting animals for meat—and I include shooting and fishing here—causes pain as well as death. It brings about, of course, a premature death—the animal dies earlier than otherwise it would. Does it bring about also an increase in pain, causing the animal to suffer more than otherwise it would? Very probably not. Overall lifetime suffering is very likely decreased. Of course, lifetime pleasure will be decreased also. So, in assessing the death, it may be of both interest and importance to know what life had in store.

Traditional Farming

I mean this to cover what is elsewhere referred to as family farming, nonintensive farming, hobby farming, organic farming, and the like.[4] Meat animals here undergo premature deaths, some pain, and some restrictions on freedom.

Factory Farming

Animals in factory farms suffer a premature death, considerable pain throughout their lives, and considerable, and discomfiting, restrictions on their

---

4. See DeGrazia (2009) for more on these distinctions, and for an argument that animals from such farms should not be eaten. It should be noted in particular that the production of organic meat reveals a concern for human health, but none especially for animal health.

freedom. I shall say almost nothing more about this. There is little point either in defending the indefensible or in attacking a practice that almost every reader here will already condemn.

Recall the opening. Why is it uncontroversial to claim that we can raise, kill, and eat vegetables? It is because nothing we can do to them, absenting side effects, is of moral concern. Shall we say that what we can do is in no way bad for them? Hardly. Killing, mutilating, confining, force-feeding, and starving plants might, in recognizable ways, be bad for them.[5] All might interfere with and impede their flourishing. But though it is bad for them, it isn't bad in a way that matters. There is no reason for us to be concerned, just for their sakes, about their well-being. Plants, we might agree, lack moral status.[6] Similarly, killing, mutilating, confining, and causing pain to animals is bad for them. Does this matter? Surely pain matters. And surely they do have moral status.[7] There are always reasons, even if defeasible, for not causing animals pain. But it is more complicated with death. This is why the focus lies here.

As we can do things that are bad for plants, so too can we do things that are good for them. They can be harmed, and they can be benefited. And, following from this, I suggest we can talk of plants living bad or good lives. Yet, as they lack moral status, it isn't bad or good in the way that matters that they live these lives. There aren't reasons, for the sake of the plant, to end their lives when their lives are bad, nor to sustain them, when their lives are good. Animals can similarly live bad or good lives. In ways that matter? Often, as I'll explain, there will be reasons to end their lives, when they are bad. But things are less clear, I'll suggest, concerning their good lives.

The Badness of Death

Farming and hunting both curtail animal lives. Is it bad, and in a way that matters, for these animals to suffer a premature death?

---

5. I claim further that in so acting we harm plants and act contrary to their interests. Against at least some strong implications in Singer (1993) I hold that non-sentient beings can have interests. See for discussion, Belshaw (2001) pp. 126–128.

6. Of course, some will disagree. See, for example, Stone (1972).

7. Or at least, animals that can feel pain have moral status. I believe all mammals and many non-mammals to be included here, but doubt whether most animals, in terms both of numbers and kinds, are, to use Richard Ryder's term, painient.

If, as Epicureans claim, it is never bad for one of us to die, then surely also it is never bad for animals. But this Epicurean View is not well argued.[8] Even if it is true that being dead isn't intrinsically bad—it won't involve you in feeling pain—it may still be relationally or comparatively bad, may still be worse for you than life. Better, then, is the Deprivation View, holding that death's badness is a function of the good life it takes from you.[9] So, when further life would have been good, then death is bad because, and to the extent that, it deprives you of this life. Though better, I claim that this view is not altogether correct. For it asks us to believe that death is bad for the one who dies, or worse than life, or worse than some other death, over a range of cases where, intuitively, this seems not to be so. It suggests that (on the assumption that future life will be good) as death at 20 is worse than death at 80, so death at 2, or at 2 weeks, or 2 days is worse still.[10] It suggests also that given a choice of six months of good life and then death, or a radical mind-altering operation followed by six years of good life and then death,[11] you should choose the latter. More generally, it seems to suggest that the badness of death is independent of your desires for and interests in further life. Plausibly, we need restrictions. And I claim:

Death is bad for you (in the way that matters) only when it cuts off a good and unfolding life that, not unreasonably, you want now, or wanted earlier, to live.

This is offered only as a necessary condition of death's badness. Death robs you of future life. But even if this life would be good, it is not bad for you not to live it, unless it relates appropriately to your life so far. It isn't bad for you

---

8. How closely the Epicurean View—death is not bad for the one who dies, and this principally because the dead either lack sentience, or existence, or both—relates to the historical Epicurus is a matter for debate. See for the revival and some recent statements of the view Nagel (1970) and various papers in Fischer (1993). For the historical Epicurus see (unsurprisingly) Epicurus (1926), Warren (2004), some of the chapters in Taylor (2013).

9. For the Deprivation View see again, Nagel (1970), Fischer's introduction (1993), and Kagan (2012).

10. See Bradley (2009) and Belshaw (2013) for discussion.

11. Suppose the operation, though allowing for a later good life, induces profound and irreversible amnesia, along with substantial character change. See, in particular, McMahan (2002) p. 77 for discussion of this sort of case.

not to live the life of a wholly different person. Nor is it bad if, though recognizably your life, you have, and have earlier had, no desire to live it.[12]

Although this amendment to the Deprivation View aligns it more closely with several of our intuitions, it has what many will see as unwelcome implications. Controversially, I claim it is not bad, in the way that matters, for babies to die. Even when, predictably, their future life will be good and, again predictably, they will at some time want very much to continue with it, this life isn't something in which they have any interest now.[13] Again, the point here concerns just the fate of those who die. The death of infants is of course very often a tragedy for the parents. And their deaths, on a wide scale, may be a tragedy for all of us.

I claim also that it is not bad, in the way that matters, for most non-human animals to die. Is this simply because, even if they can live good lives, they have no future-directed desires? Perhaps this is true but the argument needs only the weaker, and surer contention that they don't have desires for more life, or desires for that which gives them reason to want more life.[14] Squirrels, for example, don't want now to go on living so that they can eat the nuts they have stored away in the winter months. Hens don't want to survive in order to meet up soon with the farm's new cockerel.[15] Am I claiming this of all animals? No. Higher mammals have a more complex psychology, and perhaps some of these can have future concerns relevantly similar to ours.[16]

---

12. Imagine a depressed teenager who wants right now to die. Death is still bad for her. She used to want the life that, assuming recovery, she will again enjoy. If she never wanted this life then the badness of death is less clear. See for discussion Bradley (2009) pp. 128–129, and Belshaw (2013)

13. The critical term here is ambiguous in a way that is worth noting. Living on may be good for, or in the interests of, the human organism, just as it may be good for a tree. But the baby has now no desire for, or interest in, its future life.

14. In an earlier draft, and drawing on a well-known distinction made by Bernard Williams, I suggested that animals have no categorical, but only conditional desires. Yet even though it has, as I believe, some merit, I am inclined now to agree that the problems with this distinction undermine its value here. Fortunately, it isn't needed. See Williams (1973). And see for valuable discussion Broome (1999), Bradley and McDaniel (2013).

15. I don't deny that a squirrel might want to store nuts now, or (later) want to eat them now, nor do I deny that a hen might want sex now. What I deny is that such animals can want, at one time, something to happen at some future time. And (to refer again to Williams's distinction) I deny that they can want now either to stay alive in order later to eat food or have sex, or want now to eat food or have sex in the future on condition they are then alive. Nevertheless, the argument here requires only the former claim.

16. Roger Scruton, who in this area is unfairly best known for his support of fox hunting, holds that for some animals, death can be an "evil," a "calamity," a "catastrophe" (2000) pp. 45–46. Which animals? He isn't explicit, but references here to rabbits and mice suggests what might be an implausibly generous view.

But certainly it is true for most animals, and certainly for most of those that we eat.[17] These animals, I claim, are best thought of as having psychologies comprising more or less discrete series of episodes which, even if they resemble one another, are not perceived as such from the inside. And it is not a matter of regret, I contend, if this series should at any time come to an abrupt end.[18]

The Goodness of Death

As the deprivation account holds the value of death is a function of the value of the life it takes away so, as it can be bad to die, it can sometimes be good. Or at least, it can be better to die than to live. And I claim:

> Death is good for you (in the way that matters) when, but not only when, it cuts off a life that is worse than nothing, and you don't want to live this life.

Here I offer only a sufficient condition of death's goodness, allowing it might be good also in several other circumstances. It might be good for you when, for example, you believe honor requires that you die, or when it secures for you everlasting fame. It might not be good for you when for some reason you want to live a life that is worse than nothing, or don't want to live a life that will be worth living. But death will be good for you, or better than life, at least

---

17. Some people eat whales and some eat monkeys. Most of us—most readers of this book—will eat neither. Typical food animals, I claim, don't want to go on living, don't want to avoid death. Of course, many will object. I attempt to sidestep the objection toward the end of this chapter. But elsewhere I tackle it more directly. See Belshaw 'Death, Pain and Animal Life' in Visak and Garner (2015).

18. This claim about animal death is controversial. Much less controversial is the claim that death is bad for animals, even if it is considerably less bad for them to die than for human beings to die. See, for example, McMahan (1998) and DeGrazia (2009). But the dispute here is not, or should not be, real. I allow that death comes as a harm to animals, and it is bad for them to die, but, as with plants, it isn't bad in the way that matters. Further, and as I've said, I agree that animals, in virtue of feeling pain, have moral status. But it doesn't follow from this that it is at all wrong to harm them, or to kill them. (And see here Harman (2011) for suggestions to the contrary.) Those claiming that animal death is in some measure bad might agree with all this or—as it most often appears—insist that the badness gives us reasons not to kill. And in this case we are at odds. But note that the dispute can have different forms. It can be held that death is bad, and in a way that matters, either (a) personally, for the animal that dies or (b) impersonally, for the universe. Norcross (2013) p. 424 makes this distinction, and holds that (b) is true. I hold that (a) is false and that a case for (b) is yet to be made. See also footnote 22 below.

in the circumstance when it's true both that you don't want to live this life and that the content of life will be overall bad.

Death can be good, or at least better than life, for an animal. First, and uncontroversially, it is good, or better, when the alternative is only endless agony, and through all its stages a future life would be worse than nothing. Second, even if the agony will end, and the animal will live again a good life, it can be better that it dies than that the agony for now persists. Even though it doesn't want to die, it doesn't, as I've claimed, want to live this good life, it cannot, as we can, think that this agony is worth putting up with for what lies beyond. (Am I saying that the animal wants the agony to end? This might be said. But it might be said only that it is better for the animal if the agony ends, and it has no desires at odds with this).

A third claim is more controversial. Suppose the animal's life is good now and will so continue for at least a while, but that there is agony further ahead. Assume it either dies now, forfeiting the good and avoiding the agony, or it lives to experience both. So assume that painless death at the end of the good period isn't an option. Then it is better that it dies now. Again, unless its psychology rather closely resembles ours, the animal doesn't now want this future good life, and wouldn't think, as we might, that it is worth paying for later. And if death isn't bad for an animal it can, in circumstances like these, easily be good.

The appeal here isn't to some straightforward cost-benefit analysis, whereby death is better for an animal if its future life will be bad on balance. Even if there are many years of good life ahead, still it is not worth an animal's suffering a day's agony in order for it to live that life. And this is true whether the day's agony comes at the beginning or end of those good years. Even if the pleasures outweigh the pains, they can't properly compensate the animal for undergoing them. Or so I claim. Do I claim that all animals can be benefited by death? No. As it's hard to see how death can be good for a plant[19] so too for animals, if there are any, that don't feel pain. Perhaps death cannot be good for insects, worms, molluscs. But it can be, and very often will be, good for the animals that most often we eat.[20]

---

19. So while it is fairly clear how death can be bad for a plant, even if not in a way that matters—it puts an end to its flourishing—it is less clear, I think, how even in this innocuous way it can be good.

20. Most often, the meat-eating readers of this book are likely to consume chicken, pork, or beef. Even so, in terms of individual animals, they perhaps eat more shrimp, mussels, clams. In several parts of the world, insects, grubs, and snails need also to be factored in.

Killing

If it isn't bad for animals to die, does it follow that we are permitted to kill them? And if death is good are we, further, required to kill them? I make no such sweeping claims. There are very often reasons not to kill an animal—it is someone's pet, or it belongs to a rare breed, or other animals depend on it. Still, if we are thinking just of what is good for the individual animal these reasons carry no weight. A further reason, linked now to the animal's good, is that killing might cause it some pain. Though I believe it is possible to bring about a wholly painless death, this contentious view is not needed for the argument. And killing is first permitted, second required, in a range of cases, even when pain is involved.

Consider these claims in relation to veterinary practice. The idea that we should sometimes put an animal "out if its misery" is widely accepted. And though we might take extreme measures to relieve pain and extend a pet's life for the sake of its owner, it is often accepted too that if we are thinking just of the animal then death now might be best.[21] Further, and more controversially, it may be that a pain-free animal ought to be killed. Here's a plausible case. The vet detects a tumor in your cat that doesn't yet, but soon will, cause it pain. He offers to put the cat down now. Let's suppose you're planning some sea voyage, with the cat, and will be out of contact for some months. It will be difficult, in those circumstances, to do anything about its inevitable pain. You should accept the offer.

The vet will, in killing an animal, cause minimal pain. What about cases involving more than minimal pain? Shooting a badger is permitted, even when the shot isn't clean, if the badger is already wounded, and in pain, and will suffer more pain overall if it isn't shot. And shooting a healthy rabbit is permitted, I claim, even when the shot isn't clean, if, unless it is shot, the rabbit over its life will suffer equal or greater pain.[22]

In the first of these cases not only are you permitted, you are plausibly required to kill. What does a requirement depend upon? If death would be

---

21. Yet Elizabeth Harman (2011) p. 732, in discussing such a case, states, though without any argument, that surgery is permissible.

22. This isn't quite right. Assume that if you shoot the rabbit it is, before death comes, for a few moments in agony. Assume that if you don't it has some pain through its life but is never in agony—at each moment there is some compensating pleasure. Assume that it makes sense to sum the instances of pain, and that when summed the greater pain is in living. Even so, shooting isn't permitted. Agony doesn't outweigh inconvenience. This detail can, I think, usually be ignored.

good (or at least better than life) for the animal, we are thinking just of its good, and other things are equal—it would be relatively straightforward, it doesn't have young that depend on it, you are not sentimentally attached, not generally squeamish, not otherwise engaged—then, I suggest, you ought to put an end to its life.

In the second case there is no such grounding of a requirement. Death now wouldn't be good for the animal. But if death wouldn't be bad for it then (again absenting side effects) the permission remains. And if, supposing it isn't killed, inevitably it will go on to suffer a more painful death (say, when attacked by a fox, or by crows, or when succumbing to some agonizing disease), and assuming also that if it isn't killed now, then it can't be killed at all (say, this is the last time it will be within range), then killing now, even if its life is good now, is required.

Thinking just of what is best for them, of which animals is it true that we have no such reasons for killing them now? First, those that don't feel pain; second, those (if any) that have desires for a future life; third, those (and there are many) that feel no pain now, and can be killed later.

Eating

Consider the animals that we are permitted or required to kill. Are we then permitted or required to eat them?

Other things equal, I suggest, we are permitted to eat dead animals. But other things may not be equal. The permission is moot if other people, owners or onlookers, object, and moot too if other animals, hungry or grieving, have some competing claim. It is moot in a different way insofar as the animals, or their parts, are inedible—either there is no meat, or the meat is in some way tainted.

Assume the permission. It doesn't, of course, follow that you are then required to eat the meat. And only rarely will there be any such requirement. Perhaps someone might be required to eat meat on health grounds, either because of some strange idiosyncrasy of their metabolism, or because there is no other food available. There may be religious or cultural practices that involve meat-eating. Or perhaps, more fancifully, you fall into the hands of gangsters who promise not to kill some villagers if you eat hamburger. But for almost all of us, almost all of the time, eating meat is unnecessary. Those who eat it do so for pleasure or for convenience. And it isn't among the higher pleasures. No one will say you ought to enjoy eating meat, even if they say you ought to give it a try. No one will say that the lives of meat eaters, like the lives of opera lovers, go better than they would otherwise. So meat eating will be at best permitted.

Yet now there is a complication, and this may turn out to be a theoretical permission only. If you kill an animal you might as well, if so inclined, then eat the meat, but if you kill enough there will sooner or later be no more meat available. Many species have been driven to, or near to, extinction by our using them for food. Animals disappear through our greed and short-sightedness. But, of course, I've suggested that there are moral reasons also for causing them to disappear. We have, I've suggested, reason often to end the lives of existing animals. But we have also, and more plainly, reason to prevent their coming into existence in the first place. For if a life can never be in the right way good enough to make it worth continuing, and can, and most likely at some time will, be bad enough to make it worth ending, then this life is probably best not started. Certainly, many of the animals we eat, and many others besides, have lives of this kind. Aiming just at their good, and ignoring side effects, will have us be rid of them. Inevitably this will impact on our diet.

What we might call this Animal Annihilationist argument might find allies, or partial allies, elsewhere. It is, of course, more or less the game plan of many so-called Animal Liberationists, who seek not so much to give to animals a life of freedom as to prevent their being born. If they, along with vegetarians and vegans, win the argument then farm animals (sheep, goats, pigs, cows), managed animals (deer, pheasants, grouse), along with laboratory animals (rats, mice, guinea pigs) will all virtually disappear. For there just isn't a viable alternative wherein we breed and manage such animals and then provide care for them in old age.[23] And so those wanting to continue to eat meat will need to find reasons for keeping food animals in existence. What needs to be considered, then, is whether it is in any way required that there be in place the practices and procedures connected with eating meat. So suppose there are reasons for sustaining certain animal populations. But suppose also that, in spite of these reasons, there are unlikely to be these animals unless there is some benefit, or perceived benefit, direct or indirect, for human beings. For otherwise, given the various costs involved, we just might not do what, overall, we ought to do. The availability of meat is such a benefit, and might motivate us to do what overall is for the best. But why might it be good to have these animals in existence? Consider:

---

23. There are, of course, zoos. But these deal with animals in only small numbers. And to what extent their dealings are successful is, of course, disputed. See, for a recent controversial case: http://news.nationalgeographic.com/news/2014/02/140210-giraffe-copenhagen-science/. My sympathies here are very much with the protesters.

Aesthetic Reasons

Though enjoying meat isn't a part of the good life, enjoying the beauties of nature arguably is. Think about the Scottish Highlands, with mountains, heather, grouse, and deer. The image here is not of untrammeled nature but of a nineteenth-century construct, close on the heels of the clearances, Scott's novels, and Victoria's enthusiasm. Whatever its provenance, many think it important that it be sustained.[24] Curiously, some people will pay to shoot grouse and deer. Less curiously, others will pay to eat them. Insofar as we support these practices we help preserve this landscape. Without them economic pressures for a very different use of the land are harder to resist.

Environmental Reasons

The first suggestion is that meat eating is linked, and can legitimately remain linked, to some higher human pleasure. Why care about the environment? Some will hold that its value, too, is value for us, though often disguised. But others will insist it is of direct concern, and that nature matters in itself.

Can the environment look after itself? Not always in a form of which we approve. There are, near where I live, certain rare moths that feed and breed on a similarly rare balsam. These require, for their successful propagation, the ground periodically to be broken up. The most efficient and environmentally friendly way of doing this is to breed cattle that forage in the woodland.[25] This wouldn't be viable, however, if the cows didn't end up on plates.

Animal Welfare Reasons

Do we need animals for the sake of further animals? Again, to use a local example, cows, and more particularly cowpats, encourage flies, which in turn encourage swallows. More generally, meat and dairy farming is better in many respects for the wild bird population than an arable alternative. So, insofar as there is value in sustaining these populations there are reasons to foster meat

---

24. Though it should be noted that many of nature's friends push the case for significant increases in the number and kinds of wild animals here, with wolves and bears being considered for reintroduction, see http://www.bbc.co.uk/nature/22287080.

25. See http://www.telegraph.co.uk/earth/3548154/Rare-moths-saved-by-herd-of-cows.html.

consumption. Is there value here? The case for meat would be stronger if without the cow population these birds would suffer. A more likely outcome is that they disappear. So the major threat here might be one of extinction, or at least of local extinctions, rather than of reduced welfare, properly understood. Nevertheless, it is clear that a concern for some species can generate a concern for others.

In all these cases we are considering using some animals in order to bring about benefits elsewhere. And meat, as a bonus, figures here—the initial and immediate benefit will not on its own motivate us to engage in the practice, but factor in the pleasures of meat eaters and the policy becomes viable. There will be well-founded objections, however. Take the case that earlier I said would be little considered. Even if benefits outweigh costs—one dead cow can please a lot of people—many, apart from thoroughgoing consequentialists, will hold that factory farming is indefensible. Torturing one is perhaps permitted when it saves hundreds of innocent lives, but arguably indefensible for lesser benefits. And the quasi-torture of the factory farm cannot be justified no matter how many of us enjoy its products. But even if we focus on the better lives offered by traditional farming and hunting, doubts will remain. If these lives, though better, are still not worth living, how can we justify starting them in the first place? The annihilationist arguments remain in place, and the meat-eater needs to do more. One strategy is to tackle the issue raised here head-on, and argue that the benefits elsewhere outweigh the costs incurred by the animals we eat. Another is to deny these animals are, in any relevant way, harmed. And then another, more ambitious, is to argue that those that are eaten will themselves benefit. I'll say no more about the first of these strategies, but in what follows consider both the second and third.

The Value of Life

Which lives are worth living? We might, in considering this, ask also which are worth continuing, which worth starting? Human lives are often worth continuing. It is often worth paying some price, perhaps undergoing a painful operation, to sustain a life. Often, we want, and have reason, to live on. As well as this, we often, on a moment-by-moment basis, enjoy and take pleasure in our lives. They are, in this sense, often worthwhile, or worth living. Whether or not our lives are ever worth starting is more complex. None of us wants to come into existence. But even if it is not straightforwardly good for us if our lives are started, someone may think it is still a good thing, good in itself, or good for the universe, if lives like ours come to be.[26] There are, other things equal, no reasons against, and perhaps some reasons for, starting such lives.

Animal lives contrast here. These are, I've claimed, rarely worth continuing. None of those that we eat want, or have self-directed reasons, to live on. Nor are these lives, for their own sake, worth starting. No animal wants to come into existence. Are they worth living? The lot of most of them is at best dull and nondescript, and often involves a near-endless struggle just for mere survival.[27] Even if it can be good for others, it is hard to see how it can be good for these animals, or good in itself, or good for the universe that such lives are lived. It seems it may well be best, for them, that they never exist.

Yet two kinds of qualification need to be made here. First, even if animal lives are overall never worth living, and even if taken as a whole each of them will be worse than nothing, and best not started, there can be, as I've said, and especially with domestic animals, managed and early exits. The lives of both pets and farm animals in particular can be ended before they become worse than nothing. Second, there are some that very plausibly are worth living. Pet dogs appear, often, to have pretty good lives and to get various pleasures and enjoyments from, and throughout, their existence. We might think that their lives, once underway, are in a recognizable sense good for them. Even so, it isn't clear that they want to live on, isn't clear we should subject them to painful operations so they can live on, and isn't clear that it is bad for them to die. Are their lives worth starting? Assuming they are, as it appears, mostly pleasant and enjoyable, then even if it isn't good for them, it might be good in itself, good for the universe that they begin to exist. Certainly there seems to be no reason not to start such lives. And if or when there comes to be reason to end them, they can be ended. Dogs, more clearly than people, can be expected to enjoy lifelong pleasure. Certainly, then, there are reasons overall to bring some dogs into existence. It is good for us, perhaps good in itself, and either good or at worse neutral for them that they exist.

Parts of Lives

I've made already the contrast between the life taken as a whole, with its natural ending, and an earlier part, ended by us, before the bad times

---

26. I take it that Singer's (1993) suggestion that we might justify killing so long as we replace those animals killed is relevant here. But it is hard to see why someone who believes the overall level of goodness in the universe should be sustained wouldn't also think it should (other things equal) be increased.

27. Scruton (2000) p. 47 makes this point at greater length.

begin. If we consider now a still earlier part, then it may be that a case for meat can be revived.

Cats are very different from dogs. They are aloof; self-reliant; unwilling to be bullied into performing, fetching sticks, carrying newspapers; they are not anxious when left alone. Another way to view them is as dull and boring. They sit around all day, occasionally killing birds. But contrast adult cats with kittens. Kittens are a lot of fun. They appear to play, to explore, to enjoy life. You might think that the life of a kitten is worth living. There is the same contrast to be made if we consider sheep and lambs. No life appears more dreary than that of most adult sheep. They are mere eating machines, and show no interests in anything apart from food and keeping their distance.[28] But lambs have a lot of fun. Maybe their lives are worth living. There are similar distinctions, though less pronounced, between young and adult pigs and cows.

Is this utterly fanciful? Well, I make these comments after many years of regularly observing farm animals outside my home. That there are such differences, at least on the behavioral level, is undeniable. And there may be a biological explanation. Perhaps curiosity and pleasure motivates young animals to explore and learn about their environments, and so learn how best to survive. Once the lessons are learned this motivation declines.

And now the meat-eater might make something of this:

Animal Generation Reasons

Suppose the changing character of some animal lives, from young to old, is such that we are first required to bring these animals into existence in order that they live what at the outset are worthwhile lives, and then later, when their lives are no longer worth living, we are required to end them. Having ended them, there is no reason not to eat them. And suppose also that we will not do what we are required to do unless we can eat them—without meat at the end of the procedure we will, given the costs incurred, be inadequately motivated. In this picture it turns out to be good for animals that we bring them into being in order to kill and eat them.

This is too ambitious. It tries too hard to make something of the different periods in an animal's life, exaggerating the contrasts in order to legitimate our practices. We'll need to settle for less.

---

28. For recent and less recent examples of hostility to sheep see George Monbiot (2013) and Aldo Leopold (1949).

What I've suggested about young animals is that we can plausibly believe that they get considerable pleasure from, and enjoy, their lives. Nevertheless, if they lack future-directed desires, and don't want to live on, it wouldn't be wrong, even at this stage, to kill them. Nor is it clearly right to start these lives. No matter how pleasant they might be, ignoring side effects we are, I think, merely permitted, and not required, to bring these animals into existence. Still, that they enjoy their lives seems to make some difference to our permissions here.

Later in life, I've suggested, positive enjoyment ceases. But even if these lives are no longer worth living, and even if we can reliably predict there is distress and pain to come, there isn't any immediate requirement to end them. The meat-eater is involved in gross self-deception if he claims that it is best for the sheep or cows if they are killed now, even if it is better that they are killed before disease and deterioration set in. But again, that they are not even enjoying their lives seems to make some difference to the legitimacy of killing them. And remember my conceding that there may be no pain-free deaths. Killing them is legitimate only if they would otherwise suffer more pain overall.

The resulting picture, less ambitious, still allows for and perhaps encourages meat-eating. Even if, considering these animal lives as whole, and, ending in a natural death, it would be better were their lives not lived, we can, by looking to their parts and a managed exit, adopt a different view. It isn't bad for them, is perhaps good for them, and is very likely good overall when such animals have these short lives. But they wouldn't exist were it not for meat.

Animals and Humans

This account of the lives of certain food animals shouldn't seem outrageous. It significantly resembles what many of us think about human lives, where again fluctuating fortunes determine the appropriateness of certain interactions. Many of us think that even if lives are worth living, there is no requirement to start such lives. And we think that though it would be very wrong to kill someone in the prime of life it would be right to kill them when terminally ill and in great pain. Of course there are non-trivial differences here—the worthwhile portion of the average human life is much longer, and an end period where life is not worth living might not occur at all; killing in the early stages is clearly proscribed, and in the later stages, supposing it then occurs. is voluntary and often self-inflicted; we are suspicious of, rather than looking for, motivation in terms of personal gain. But it isn't clear that these differences, though several and important, are such as to upset the basic comparison.

Comparisons can be taken further, however. And it might already have been noted that parts of the argument here have affinities with the anti-natalist views of David Benatar.[29] Whereas I've suggested that animals might have lives that are worse than nothing, his claim is that this is also, and almost always, true of human lives. We both put more emphasis on the badness of pain than some others might, and we both explore, though each in different ways, the difficulties of countering pain with pleasure. Both of us, though perhaps to different degrees, are willing to acknowledge the counter-intuitiveness of parts of our positions.

This isn't the place to engage further with Benatar's argument, or to explore its alleged weaknesses.[30] What I do want to do, however, is explain how a modified version of his anti-natalism provides an inversion, or mirror image, of the meat-generating argument offered here.

Benatar's central contention, that our lives, because of the pain that almost inevitably figures within them, are worse than nothing, can, I think, be given in a defensible guise, at least for parts of our lives. But I want to make here not just the familiar point that some lives, especially in their closing stages, might not be worth living, and might best be ended, but also that all our lives, in their opening stages, are like this. What I've said of animals might be said also of babies. A baby's life is perhaps a mix of trivial pleasures and more serious pains—being born, colic, teething—and a mix, too, where the good, even if it outweighs, cannot compensate for the bad. A baby cannot think its pain will soon end, has only recently begun, that pleasure is on its way, or that pleasure is (if indeed it is) the norm. It cannot think that pain might bring benefits, or that it might be worth undergoing as the price for future life. Its distress, when it is distressed, is its entire world. Better for a baby, perhaps, never to be. Or at least to grow out of it.[31] But should we think that as they grow out of it so later pleasures compensate for earlier pains? Arguably not. For infant and adult psychologies, we might think, are so tenuously connected as to prevent their

---

29. Benatar (1997) and (2006).

30. More important, I allow, while Benatar denies, that life might be worth living at a time when it contains some pain, if at that time there is outweighing pleasure.

31. Imagine doctors tell a would-be mother that any child she has will live only for a week, seemingly normally, and then die a painless death. Or, for a larger scale example, that an asteroid will hit this planet nine months and one week from now. There seems to me to be here reasons not to conceive. These short lives are not worth living.

being seen as belonging to the same person. Indeed the baby isn't a person at all, and has neither an interest in nor concerns for the person to come.

The mirror image, then, is this: human beings, I am suggesting, begin their lives with a phase that is worse than nothing but soon, with normal development, and the onset of personhood, enter an extended period where life is well worth living. We might hold that it is wrong to start such a life but, if started nevertheless, it soon becomes wrong to end it. Animals, in contrast, or at least some of them, start out with the good times, but then all too quickly move on to tedium. So it is right, or at least permitted, to start these lives, but then soon right, or permitted, to end them.

Again the differences can be explained. We are born, as in a way we must be,[32] in some sense prematurely, helpless, unable to fend for ourselves for some considerable time. It isn't surprising that at the outset we find life hard. Animals, in contrast, are very soon able, more or less, to make their own way. It is as if they are almost immediately into childhood.

Objections and Replies

Although there's been much discussion here of animal welfare, the question of whether these creatures have any rights, and in particular a right not to be killed, has, as will have been evident, thus far been overlooked. Suppose I were to say, as, admittedly, might be tempting, that I don't believe that animals have such a right. It could be objected that this is a matter of vital importance, and one should err on the side of caution. For it may be that what we do, even if we kill animals altogether painlessly, is akin to murder.[33] And so long as such a possibility cannot be altogether excluded, then there can be no defense of even the most benign forms of hunting and farming.

Would this be taking caution too far? There might be two underlying claims here—first that animals, in spite of appearances, have psychologies similar enough to ours for them to fall under the same moral umbrella, and second, that their right to life doesn't at all depend on their psychological makeup.[34] But, to begin with this, it is perhaps not merely difficult, but rather deeply impossible to believe that there might be rights to life, or to continued

---

32. See http://www.scientificamerican.com/article/human-babies-long-to-walk/.

33. See Rachels (2011), where this "Argument from Caution" is described as "excellent."

34. A middle position here might be that their having such a right derives from their ability to feel pain. So psychology matters, but further similarities take a back seat.

existence, which are enjoyed by non-sentient animals, or plants, or inanimate objects in any sense that erect constraints on our behavior. Psychology must have something to do with it. Suppose, then, that my claims about this are more or less correct, and the best way to construe many animal minds is as a discrete series of episodes, most of them containing some instances of pleasure and/or pain, but none of them cohering into anything remotely resembling what, in the human case, constitutes an ongoing adult life. Could it really be that something's rights are violated, if this series is terminated, or that causing such termination might be akin to murder? Perhaps it's hard to see how this can be a real possibility. Suppose, though, that I am wrong about animal psychology, and it is more complex, with more of a temporal reach, than is suggested here. There are two points to be made in reply. First, what I am surely not wrong about is in ordering the room of skepticism—it is far more likely that cetaceans, apes, and dogs are in some sense persons or quasi-persons than are rabbits, sardines, frogs. The "higher" the form of life, the more likely that killing is in some way like murder, and the more appropriate is caution. Second, wholesale resistance to my claims about animal psychologies (and there will be such) suggests the following compromise: view the argument as one about hypothetical beings, and append to it some queries about the real world. I suggest, that is, that there could be creatures whose psychologies are such as described here—they feel pain, but are deeply fragmented. We can then ask: Would death, killing, eating, be bad for these creatures, and if so, in what ways? Then we can further ask, having done with philosophy, to what extent these creatures resemble the animals we know.

There is suggested here, however, a further and more interesting objection. Insofar as an animal life is fragmented, and comprises a series of distinct stages rather than one unfolding whole, then it is perhaps unclear how we can legitimately intervene at one stage in order to produce a benefit in another. Perhaps I can harm you now in order to benefit you later, but it is less clear I can harm one to benefit another. A key suggestion, however, has been that it might be right to end an animal's life, even when that life is going well, in order to prevent its suffering future pain. Yet if to attend to this later stage is in effect to consider another life, then the earlier intervention is suspect.

Three responses might be offered. First, I might pull back from the stage view, holding that there really is here just the one unfolding life, but one in which the subject never has an interest in the future. And this is enough to render morally neutral the ending of that life. Second, I might be more cavalier about weighing together different lives, insisting that we can indeed kill one person to save, or even to aid, others. But a third response is perhaps better. If I can't kill you to save others, that is surely in part because you want to live on. But, I've suggested, animals don't want this. So even if the different stages in an animal's life should be seen, in effect, as distinct lives, there is no obstacle to ending one stage to prevent a later stage from starting. Or if the stages view is now taken further, there is no reason not to terminate an earlier series of good stages in order to prevent there being a later series of bad stages. And as there is no reason not to do this, so there is reason to do it.

Summary

Are we permitted to eat animals? The argument has been that as we're permitted to kill them so we might then eat them. And killing is permissible as, even if they can lead good lives, these are not lives that they want to live. Are we required to eat animals? No argument has been offered for this. I've suggested, however, that we might be required to bring animals into existence, but that we're in fact unlikely to do this unless adequately motivated. And meat motivates. Yet there is a puzzle here. Is it at all plausible to suppose we might be required to bring some things into existence and at the same time permitted to kill them? As I've suggested, someone might think that we should create animals, for their sakes, just when they will lead good lives. Think this, and it is hard to see how we might be permitted to kill the very same animals, unless there is some countervailing reason in play. But I've made no such claim, suggesting only that we are permitted, when the life will be good, to start animal lives, and we are perhaps required to start them when other factors—for example our pleasure at there being such lives—are added to the account. And this is, I've argued, consistent with a permission to end the lives. Yet suppose the stronger claim is in place, and that we are required to start good animal lives for the sake of these animals. It can still be held that we are permitted, or even required, to end these lives when circumstances change, and they stop being good, or start to be bad.

References

Beauchamp, T., and Frey, R., eds. (2011) *The Oxford Handbook of Animal Ethics*. Oxford: Oxford University Press.

Belshaw, C. (2001) *Environmental Philosophy: Reason, Nature and Human Concern*. Chesham: Acumen.

Belshaw, C. (2013) "Death, Value and Desire" in Bradley et al.

Belshaw, C. (forthcoming) "Death, Pain and Animal Life" in Višak and Garner (forthcoming).

Benatar, D. (1997) "Why Is It Better Never to Come into Existence," *American Philosophical Quarterly* 34(3): 345–355.

Benatar, D. (2006) *Better Never to Have Been: The Harm of Coming into Existence.* Oxford: Oxford University Press.

Bradley, B. (2009) *Well-Being and Death.* Oxford: Oxford University Press.

Bradley, B., Feldman, F., Johansson, J., eds. (2013) *The Oxford Handbook of Philosophy of Death.* Oxford: Oxford University Press.

Bradley, B., and McDaniel, K. (2013) "Death and Desires" in Taylor (2013).

Broome, J. (1999) "The Value of a Person" in *Ethics out of Economics.* Cambridge: Cambridge University Press.

DeGrazia, D. (2009) "Moral Vegetarianism from a Very Broad Basis," *Journal of Moral Philosophy* 6: 143–165.

Epicurus (1926) *Letter to Menoeceus,* in Bailey, C. (trans.) (1926) Epicurus, *The Extant Remains.* Oxford: The Clarendon Press.

Fehige, C., and Wessels, U., eds. (1998) *Preferences.* Berlin: Walter de Gruyter.

Fischer, J., ed. (1993) *The Metaphysics of Death.* Stanford: Stanford University Press.

Harman, E. (2011) "The Moral Significance of Animal Pain and Animal Death" in Beauchamp and Frey (2011).

Kagan, S. (2012) *Death.* New Haven, CT: Open Yale Courses.

Leopold, A. (1949) *A Sand County Almanac.* New York: Oxford University Press.

McMahan, J. (1998). "Preferences, Death and the Ethics of Killing" in Fehige and Wessels.

McMahan, J. (2002). *The Ethics of Killing.* Oxford: Oxford University Press.

Monbiot, G. (2013) *Feral: Searching for Enchantment on the Frontiers of Rewilding.* London: Allen Lane.

Nagel, T. (1970) "Death," *Nous* 4.1: 73–80. and reprinted in Nagel, T. (1979) *Mortal Questions.* Cambridge: Cambridge University Press.

Norcross, A. (2013) "The Significance of Death for Animals" in Taylor (2013).

Rachels, S. (2011) "Vegetarianism" in Beauchamp and Frey (2011).

Scruton, R. (2000) *Animal Rights and Wrongs.* London: Metro.

Singer, P. (1993) *Practical Ethics* (2nd Edition). Cambridge: Cambridge University Press.

Stone, C. (1972) "Should Trees Have Standing? Towards Legal Rights for Natural Objects," *Southern California Law Review* 45: 450–501.

Taylor, J., ed. (2013) *The Metaphysics and Ethics of Death.* Oxford: Oxford University Press.

Višak, T., and Garner, R. (2015) *Animal Ethics: Life, Death and Welfare.* Oxford: Oxford University Press.

Warren, J. (2004) *Facing Death: Epicurus and His Critics.* Oxford: The Clarendon Press.

Williams, B. (1973) "The Makropulos Case: Reflections on the Tedium of Immortality" in *Problems of the Self: Philosophical Papers 1956–1972.* Cambridge: Cambridge University Press.

2 STRICT VEGETARIANISM IS IMMORAL

Donald W. Bruckner

Introduction

The most popular and convincing arguments for the claim that vegetarianism is morally obligatory focus on the extensive, unnecessary harm done to *animals* and to the *environment* by raising animals industrially in confinement conditions (factory farming). I outline the strongest versions of these arguments. I grant that it follows from their central premises that purchasing and consuming factory-farmed meat is immoral. The arguments fail, however, to establish that strict vegetarianism is obligatory because they falsely assume that eating vegetables is the only alternative to eating factory-farmed meat that avoids the harms of factory farming. I show that these arguments not only fail to establish that strict vegetarianism is morally obligatory, but that the very premises of the arguments imply that eating some (non-factory-farmed) meat rather than only vegetables is morally obligatory. Therefore, if the central premises of these usual arguments are true, then strict vegetarianism is immoral.

The Factory Harm Argument

The first argument for vegetarianism to consider focuses on the harm done to animals raised on factory farms. I take as my point of departure an agreement between me and the strict vegetarian on premise (P1) Factory farming causes extensive harm to animals. I use "extensive" in the double sense of large in both scope and severity. I take this point about extensive harm as so well established by the scientific and the philosophical literatures, as well as by popular media accounts, that we can simply treat it as common knowledge and a shared assumption.[1] To review just a few of the harmful practices and conditions: To reduce the natural impulse of laying hens to peck at each other that is exacerbated by their stocking density in cramped battery cages, the first quarter or third of their beaks is painfully cut off when they are young. Meat chickens are raised in enclosures housing tens of thousands of birds, where ammonia levels are so high that many suffer from chronic respiratory disease. Pregnant pigs are kept in gestation crates so small that they can barely move. Male pigs and beef cattle are castrated without anesthesia. Pigs and cattle on the way to the slaughterhouse are packed on trucks without protection from the elements and without food or water. Many suffer, and some die on the way. At the slaughterhouse, cattle and pigs are sometimes shackled and hoisted fully conscious and kicking, before their throats are cut. These would all seem to be harms, whether, as David DeGrazia (2009, p. 153) points out, one's account of harm is based on pain, an inability to engage in species-specific functioning, or something else.

Consider now the next premise: (P2) This harm is unnecessary. It is unnecessary in the sense that we humans do not need to cause it in order to gain adequate nutrition for healthy bodies and to preserve our lives. Simply put, there are other, readily available, perfectly nutritious, non-animal sources of food. I will not dwell on this point at all, except to say that it is commonly accepted in the field of nutrition on the basis of careful scientific study.[2]

Combining these first two premises, we get the intermediate conclusion (C1) The practice of factory farming causes extensive, unnecessary harm to animals.

The next premise of the argument is (P3) It is wrong (knowingly) to cause, or support practices that cause, extensive, unnecessary harm to animals. DeGrazia (2009, p. 159) argues that this premise—indeed, the whole argument I am in the process of sketching—is consistent with a variety of normative ethical theories.[3]

---

1. For documentation supporting the following claims, see Rachels (2011), DeGrazia (2009), Singer and Mason (2006), and Rollin (1995).

2. On this point, see Rachels (2011, pp. 892–893), Engel (2011, pp. 354–355), and DeGrazia (2009, pp. 154–155).

3. Pre-dating DeGrazia (2009), Engel (2000) and Curnutt (1997) similarly argue that vegetarianism is obligatory independent of the truth of any specific moral theory.

One need not be a consequentialist, for example, to accept it. I grant this crucial assumption for the sake of the argument. The next premise is (P4) Purchasing and consuming meat originating on factory farms supports the practice of factory farming. Therefore, we get the conclusion (C2) Purchasing and consuming meat from factory farms is wrong.

Philosophers have raised questions about some of these assumptions.[4] I do not know whether the questions can be met, but I do think that this general line of argument, or something like it, lies behind the most common and strongest case that can be made against purchasing and consuming factory-farmed meat. So I shall just assume for the sake of my purposes here that this argument does, indeed, establish that purchasing and consuming meat from factory-farmed animals is wrong. What I wish to question is the next step that is often taken. This is the step from the conclusion that purchasing and consuming meat from factory farms is wrong to the obligatoriness of a vegetarian diet. It is apparently just supposed to be obvious that we are morally obligated to eat vegetables rather than factory-farmed animal products because eating vegetables does not harm factory-farmed animals.[5]

Let us call the argument just rehearsed that starts from the extensive harm done to factory-farmed animals and that ends with the conclusion that we are morally required to be vegetarians—let us call this argument the Factory Harm Argument. I now show that this last step of the Factory Harm Argument is a misstep.[6]

---

4. (P3), (P4), and related claims have attracted considerable attention lately, for it is questionable whether any individual's purchase—much less mere consumption—of factory-farmed meat has any causal role in harming animals. Rachels (2011, p. 886) and Norcross (2004, pp. 121–122) recognize this causal impotence problem and argue that there is a tiny chance that an individual's purchasing and consumption decisions will have a massive impact on the number of meat animals produced. This, Norcross says, is "morally and mathematically equivalent to the certainty of saving" a small number of meat animals (2004, p. 233). Budolfson's contribution to this volume attempts to counter this sort of argument. DeGrazia (2009, pp. 157–159) recognizes the need to bridge the gap between the wrongness of factory farming and the obligations of individual consumers. This explains why he settles on a principle similar to the one reflected in (P3) that it is wrong to cause, or (financially) support practices that cause, extensive, unnecessary harm. He thinks this sort of formulation is consistent with both consequentialist and non-consequentialist moral theories and reflects our considered view that "complicity matters" (2009, p. 159).

5. Making a similar leap in a different argument, Rachels claims that "killing animals to obtain food is wrong" and concludes on this basis that "we should be vegetarians" (2011, p. 894). In fairness to DeGrazia, he does consider non-factory farms, and expresses some reservation (albeit mild) that his argument establishes the wrongness of eating meat from all such farms.

6. A final note before getting on to the positive part of my argument: I ignore the terminological distinctions among *vegetarian*, *vegan*, and *ovo-lacto vegetarian* because they do not affect the substance of my argument.

Factory Farming and Roadkill

I am going to argue, in effect, that the Factory Harm Argument presents a false dilemma. The argument assumes that the only alternative to eating meat from factory-farmed animals is to eat vegetables. That is just not the case, though, for there are alternatives to eating factory animals that avoid some or all of the harm to animals associated with factory farming. These are: (1) Meat from humanely raised and slaughtered animals; (2) Meat from hunted wild animals; (3) Meat from animals killed by vehicular collisions, that is, roadkill. I will not address humane meat or hunted meat[7] because (a) I have partially addressed them elsewhere (Bruckner, 2007) and (b) I think their cases are less clear than the case of roadkill, because only eating roadkill completely avoids supporting practices that cause harm to animals, whereas eating humane meat or hunted meat still supports practices that cause at least some harm to some of the animals from which the meat is taken.

The premises of the Factory Harm Argument, I claim, support eating road-kill at least as much as they support eating vegetables. The Factory Harm Argument appeals to the extensive, unnecessary harm done to factory-farmed animals. We are obligated not to purchase and consume such meat because doing that supports practices that cause extensive, unnecessary harm to animals. So we are obligated to eat something else. Vegetables are something else. But so is roadkill. So the Factory Harm Argument supports eating vegetables and it supports eating roadkill, since eating both avoids supporting factory farms.

I need to head off some immediate objections by clarifying the sort of roadkill under discussion and explaining some of the freegan practices involved in its collection and consumption. Under discussion is the collection and consumption of large, intact, fresh, and unspoiled animals such as deer, moose, and elk. Most US states allow individuals to collect and consume such animals, usually after adhering to some reporting requirements to protect

---

7. I acknowledge a possible problem with constructions such as "hunted meat" and "venison" versus "muscle tissue from hunted wild animals" and "deer body parts." These constructions risk either thingifying the animals by identifying them with their products or masking reality by making the animals what Carol Adams (1990, pp. 47–48, p. 67 *et passim*) calls "absent referents" when we refer to such things as steak, hamburger, roasts, and bacon rather than the animals from which these parts come. My excuse is mainly expository economy. I am also not persuaded that there is *always* a problem with language that makes referents absent or otherwise masks reality, though surely there *often* is. We also, for instance, talk about pasta sauce and coleslaw rather than tomatoes and cabbage, not to mention the poorly treated migrant workers who harvested the vegetables. The absence of the referents or in any case less-than-full picture of reality in the shorthand does not necessarily show disrespect for the animals any more than for the migrant workers. Thanks to Mylan Engel for discussion on this point.

against poachers claiming that illegally taken game was road-killed. In some states, charitable organizations and government agencies have pioneered systems for collecting, butchering, and distributing road-killed meat to needy individuals. Not under discussion is the rotting squirrel or rabbit carcass that is being picked apart by vultures, or the mangled deer spread over several lanes of highway that is conjured in many minds when mentioning roadkill.

In this connection, here are some very quick statistics. State Farm Insurance (2011) estimates that in fiscal year 2010–2011 in the United States, there were 1,063,732 claims that involved deer, elk, or moose to all automobile insurance companies.[8] Now "[e]stimates of the proportion of deer that are hit on roadways and go undocumented, and hence unreported, range from 50% . . . to more than six times the reported number" (Forman et al., 2003, p. 118). To err on the conservative side, let us just double the State Farm estimate, and take 2.1 million as a reasonable estimate of the number of deer, elk, and moose killed by vehicular traffic in the United States each year. Although elk and moose are considerably larger (and yield more meat) than deer, let us suppose that all of those animals were deer, that 75% of those deer were suitable for consumption, and that 75% of the meat on each deer was undamaged. Still estimating conservatively, a deer yields 35 pounds of meat (Pennsylvania Game Commission, 2014). So that means that those 2.1 million deer would yield approximately 41,343,750 pounds of perfectly nutritious meat. One beef animal yields about 516 pounds of meat (Iowa State University, 2009, p. 15). So if those deer, elk, and moose (collectively "venison") were collected for consumption, that would be equivalent to approximately 80,124 beef animals or 8,268,750 five-pound chickens per year.[9]

---

8. This estimate is consistent with others from 720,000 to 1.5 million (Forman et al., 2003, p. 118).

9. Objection: This only amounts to about 0.24% of the over 33 million cattle slaughtered in the United States each year (USDA Economic Research Service, 2014a) and 0.10% of the 8.6 billion chickens slaughtered in the United States each year (USDA Economic Research Service, 2014b), barely a drop in the bucket. Reply: I am not claiming that this volume is large in relation to current meat production. I am not even claiming that any animal suffering would be prevented by widespread roadkill consumption, due to the issues pointed out in note 4. Note, however, that estimates of per capita meat consumption in the United States range from 128g/day (Daniel et al., 2011, p. 579) to 212g/day (Wang and Beydoun, 2009, p. 623), or between 0.28 and 0.47 pounds per day. Taking the average of the estimates (0.375 pounds/day), those 41,343,750 pounds of meat would be equivalent to about 300,000 meat-eaters leaving the market. Although this number of meat-eaters is small in relation to the total number of meat eaters, surely no one advocating for vegetarianism on moral grounds should sneeze at any proposal that would be equivalent in this way to increasing the number of vegetarians by 300,000.

Roadkill and Vegetables

So much for that. I claim to have shown that collecting and consuming road-kill is at least as well supported by the Factory Harm Argument as purchasing and consuming vegetables, since both refrain from supporting factory farm-ing. I now argue that if the premises of the Factory Harm Argument are true, then we are positively morally obligated *not* to have diets consisting of all veg-etables. Instead, we are obligated to get some of our protein from roadkill.

There is a questionable argument against strict vegetarianism and in favor of the consumption of large pasture-raised herbivores that was put forward in the philosophical literature by animal science researcher Steven Davis (2003). Davis claims that the harm done to wild animals through raising vegetables is greater than the combined harm to wild and domestic animals through rais-ing large herbivores on pasture. His basic idea is that in the farming of common vegetable crops such as wheat, corn, soybeans, and rice, many field animals are injured or killed when the fields are plowed and when the crops are harvested. These include rabbits, field mice, ground-nesting birds, even wild turkeys and numerous amphibians (Davis, 2003, p. 389). For example, cutting a wheat field allegedly results in chopping up about half of the rabbits in the field and almost all of the other small mammals, ground birds, and reptiles (ibid.). On the other hand, using the same amount of land to graze beef cattle certainly results in the deaths of those cattle as well as some field animals. In the cattle-grazing operation, however, it is unnecessary to perform operations on the pasture with as much frequency and vigor as is needed to plow, disc, plant, cultivate, and harvest vegetable fields. So, raising large rumi-nants on pasture will result in fewer animal deaths per acre than growing veg-etables. Therefore, this argument concludes, we may be morally obligated *not* to have diets consisting only of vegetables, but to include some meat from large pastured ruminants in our diets as well (Davis, 2003, p. 393).

The nice thing about this argument is that it uses the same harm principle as the Factory Harm Argument, namely, (P3) It is wrong (knowingly) to cause, or support practices that cause, extensive, unnecessary harm to animals. Maintaining an exclusively vegetarian diet supports a practice that causes ex-tensive harm. The harm is unnecessary because pastured beef is an alternative food source, the production of which causes less harm. Therefore, it is wrong to maintain an exclusively vegetarian diet.

One less nice thing about this argument, as argued by Andy Lamey (2007), is that it relies on some questionable empirical claims to support the calcula-tions Davis uses to argue that the total number of deaths caused by a vegetar-ian diet is greater than the total number of deaths caused by a diet of mostly vegetables and some pasture-raised beef. Lamey does not dispute that a large number of animals are killed in vegetable production, but only that it is not clear that enough are killed to support the calculations that Davis uses to show that vegetable farming causes more animal harm than cattle grazing.

So, it is not clear that Davis's argument for the obligatoriness of eating some beef is successful, but I have a variation on his argument to propose. Everyone seems to agree that extensive harm is done to animals in the production of vegetables. If only we could find a source of food that did not harm any animals at all, then we would have a knock-down argument for the obligatoriness of eating that kind of food rather than vegetables, because otherwise one supports a practice that causes animals extensive, unnecessary harm, which is wrong, according to (P3). I have it: Picking up road-killed animals does not harm any animals. Road-killed animals are already dead, so *they* are not harmed by picking them up as livestock *are* harmed by common husbandry practices. And no animals are killed in the *process* of picking up roadkill, as field animals are killed in the process of crop farming. So on the very standards of the Factory Harm Argument, we are obligated to collect and consume roadkill.

To recapitulate: Factory farming causes extensive, unnecessary harm to animals. By (P3), supporting a practice that causes extensive, unnecessary harm to animals is wrong. So, instead of factory-farmed meat, the argument alleges, we should purchase and consume only vegetables, because that avoids supporting a practice that harms factory-farmed animals.

I objected to this argument on the grounds that eating vegetables is not the only way to avoid supporting a practice that harms factory-farmed animals. Eating roadkill also avoids supporting factory farming. So the conclusion that we are morally obligated to have an exclusively vegetarian diet does not follow.

I just argued that, if the harm principle is true that was used to support the claim that eating factory-farmed meat is wrong, then replacing some vegetables with meat from roadkill is morally obligatory. Once we see that roadkill is a harm-free source of food and that vegetables are not harm free, we see that the reasons usually given for strict vegetarianism support an obligation not to be strict vegetarians but to eat some roadkill.

Objection: Straw Person

Is it not, one might object, really only a straw person who would claim that we are prohibited from eating all meat under all circumstances? What if the philosophers whose view I am addressing responded that they only intend their views to apply to the purchase and consumption of factory-farmed meat and that they do not mean to prohibit the consumption of meat altogether? Indeed, there is decisive evidence for interpreting them as not claiming that the total abstinence from meat is a moral requirement. David DeGrazia, for example, is explicit that his argument does not focus "on the consumption of animal products per se" (2009, p. 148). Stuart Rachels, in his argument that most resembles the Factory Harm Argument, is clear that his "is not an argument for vegetarianism" (2011, p. 883).[10] We might see our way to excusing them for describing their views as "vegetarian." For they are assuming, perhaps, that for the vast majority of people (in many industrialized countries, at least), the vast majority of meat available to them for consumption is factory-farmed meat, so for most people (same qualification) abstaining from factory-farmed meat practically amounts to vegetarianism.

In further fairness to the philosophers under discussion, some indeed do recognize that it is perfectly permissible by their lights to eat roadkill and other meat as long as one does not support practices that cause extensive, unnecessary harm to animals. Peter Singer has been asked repeatedly in interviews about eating roadkill. In 2009 he said he does not "think there's any problem with [eating] roadkill" (Dawkins, 2009, 19:00–19:35).[11] David DeGrazia writes in a footnote to the paper from which I have been drawing that his "position does not oppose, say, the consumption of a dead animal one finds in the woods" (2009, p. 148, fn. 14). Stuart Rachels similarly writes: "Perhaps you shouldn't support the meat industry by buying its products, but if someone else is about to throw food away, you might as well eat it" (2011, p. 883). Jordan Curnutt is also explicit that his Factory-Harm-type argument "allows . . . eating animals who died due to accidents" (1997, p. 156).

So on their view, there is not "any problem" with eating roadkill; their view "allows" and "does not oppose" eating an already-dead wild animal; and on their view you "might as well eat it." So eating it is *permissible*. The concern I have with these ways of putting it is that they make it sound like eating these things is *merely* permissible: It is not forbidden by morality to do it, but it is not obligatory either. You can do it or not. From the standpoint of morality, it is a don't-care decision, similar to the decision whether to tie your left shoe

---

10. He is not always as careful—see note 5.

11. Singer makes similar claims in (Kendall, 2011 and Denton, 2004). Singer and Mason also say that eating meat from dumpster diving is "impeccably consequentialist" (2006, p. 268) and that "it is difficult to see any objection to eating meat taken from" animals killed (by hunting) to protect the environment (2006, pp. 259–260). See the next paragraph for my objection to these ways of putting it.

first or your right shoe. But clearly the strength of the premises of the Factory Harm Argument—and these are four philosophers who endorse arguments very much like the Factory Harm Argument—clearly the strength of their own premises makes it obligatory, and not merely permissible. For as I have taken pains to argue, if you fail to eat the already-dead animal and you purchase vegetables instead, then you are supporting a practice that causes extensive harm to animals that is unnecessary, in violation of (P3).

So if the position I was addressing were the position that no meat-eating is permissible under any circumstances, I could justly be accused of attacking a straw person, because these philosophers explicitly disavow that position.[12] But I am not addressing that view. I am addressing their failure to make what I have argued is an obvious inference to which they are committed by their own premises. This is the inference that it is not only permissible in some circumstances to eat meat, but it is obligatory.

The Environmental Harm Argument

Set aside the Factory Harm Argument and considerations of animal welfare. The second argument for vegetarianism that I wish to examine is based on harm done to the environment by raising livestock for food. My discussion of this Environmental Harm Argument will have the same structure as my discussion of the Factory Harm Argument.

The first premise of the argument is (P1) Factory farming causes extensive (i.e., large in scope and severity) harm to the environment. Again, these harms are well established, so I travel quickly and only mention a few.[13] One sort of harm caused by factory farming is pollution. Factory farming produces (a) CO_2, an environmentally harmful greenhouse gas, largely through operating petroleum-powered equipment. The animals themselves emit vast quantities of (b) methane and nitrous oxide, two other greenhouse gasses. They also produce (c) manure, which contaminates water when improperly managed. Growing crops to feed livestock produces (d) nitrogen runoff from fertilizer, which also pollutes water. A second sort of environmental harm caused by factory farming is the overconsumption of natural resources. This overconsumption is due to the fact that raising crops to feed animals that are then fed

12. I say that it would be a just accusation, not that I would be guilty, for Cora Diamond seems to hold the view that eating meat under any circumstances would be wrong, as in her case of the cow struck by lightning (1978, p. 468).

13. See Rachels (2011) and Fox (2000) for documentation supporting the following claims.

to humans is much less efficient than raising crops to feed humans directly. This inefficiency results in the overuse of petroleum, water, and land.

The next premise of the argument is (P2) These harms are unnecessary. They are unnecessary because we humans do not need to eat meat. This is the same premise with the same support as in the Factory Harm Argument.

Combining (P1) and (P2), we get (C1) The practice of factory farming causes extensive, unnecessary harm to the environment.

The next premise in the argument is (P3) It is wrong (knowingly) to cause, or support practices that cause, extensive, unnecessary harm to the environment. This premise is significantly weaker than the one Michael Allen Fox uses in his version of this argument: "[W]e ought to minimize the harmful impact of our lives . . . on the biosphere" (2000, p. 166). I grant the crucial (P3) for the sake of the argument. Since (P4) Purchasing and consuming meat originating on factory farms supports the practice of factory farming, we get the conclusion (C2) Purchasing and consuming meat from factory farms is wrong. Therefore, vegetarianism is obligatory.

Surely the reader can see what is coming. There is an inferential leap from (C2) to the obligatoriness of vegetarianism. Fox at least tries to support the inference by considering meat from animals raised on non-factory farms before claiming that "vegetarians . . . are able to live even more lightly on the land than do meat-eaters of any description" (2000, p. 166). That is false, though, because consumers of roadkill tread yet more lightly on the land than do vegetarians.[14]

To make out this last claim, consider pollution first. Farming crops for humans produces (a) some CO_2 through petroleum-operated farm equipment. Ideally, the collection of roadkill increases petroleum consumption only by the tiniest amount of additional fuel one uses transporting the additional weight of the deer to one's home. In the non-ideal case, one burns gasoline to drop it off and pick it up at a butcher shop. It would be hard to estimate whether this would use more or less petroleum than needed for the production and transport of a nutritionally equivalent amount of vegetables. So the best we can say about the production of CO_2 from petroleum use is that in the ideal case collecting roadkill likely produces less CO_2 than vegetable farming. As far as (b) methane and nitrous oxide go, roadkill is vastly superior if rice is among the crops produced for human consumption, as rice

---

14. Another reason it is false is that according to one study (Vieux et al., 2013), plant-based diets are associated with higher greenhouse gas emissions than some diets containing meat, because fruits and vegetables have lower caloric content per unit weight (p. 576). So a plant-based diet is not more environmentally friendly in all respects than a diet with meat.

paddies produce large quantities of methane (Neue, 1993); otherwise, road-kill and vegetables are probably even. Roadkill and vegetables are definitely even with regard to (c) manure, as neither produces any. Finally, (d) no nitrogen is put on the soil in order to produce roadkill, but it is used for crops. In the ideal case of crop production, however, nitrogen need not be added to the soil artificially, and if it is, technology exists to collect much of the nitrogen runoff before it enters rivers and streams (Blanco-Canqui et al., 2004). So in the ideal case of vegetable production, roadkill collection and vegetable production appear to be even with regard to nitrogen runoff. Considering (a)–(d) and ideal practices, therefore, roadkill collection probably produces less pollution than crop farming, but almost certainly no more.

Consider the second sort of environmental harm canvassed, the overconsumption of petroleum, water, and land. We found that roadkill may have a slight edge regarding petroleum use. Clearly no water and land need to be devoted to producing roadkill, whereas lots and lots of water and land are needed to grow crops. So from the standpoint of the use of these natural resources, roadkill has a decided advantage.

Again, the argument goes by appeal to the core premise of the Environmental Harm Argument, (P3) It is wrong to cause, or support practices that cause, extensive, unnecessary harm to the environment. We observe that purchasing and consuming vegetables supports vegetable farming, a practice that causes extensive harm to the environment. We have just seen that some of the environmental harms of vegetable farming are unnecessary because roadkill is a less harmful alternative. Therefore, if (P3) is true, then it is wrong to eat only vegetables and we are obligated to collect and consume some roadkill.

Objections

Causal Impotence

Objection: Your argument only succeeds if you assume that the practice of collecting and consuming roadkill will be causally efficacious in reducing factory-farmed animal production and vegetable production, thereby reducing harms to farm animals, wild animals, and the environment. Yet individual consumption decisions do not have such causal efficacy. If a person collects even 50 pounds of road-killed meat and abstains, therefore, from purchasing 50 pounds of meat or a nutritionally equivalent amount of vegetables from the grocery store, no fewer meat animals or vegetables will be produced. Even collective consumption decisions would not affect animal or vegetable production here, since the amount of available roadkill is so tiny in relation to the aggregate nutritional needs of populations where roadkill is available.[15]

Reply: The causal link between food consumption decisions and food production decisions is controversial.[16] So I was careful to present my argument—as well as the standard arguments for vegetarianism—in a way that succeeds whether there is a causal link or not. Recall that the principles numbered (P3) say that it is wrong to cause, *or support practices that cause*, extensive, unnecessary harm to animals or the environment. So if there is a causal link between collecting and consuming roadkill and reducing harm, principles (P3) say it is wrong not to collect and consume roadkill, because otherwise one causes unnecessary harm. If there is not such a causal link, the principles still say it is wrong not to collect and consume roadkill because purchasing other food *supports practices that cause* extensive, unnecessary harm. In either case, it is wrong not to collect and consume roadkill.

The idea behind the second branch of the statement of the principles is that complicity matters. Participating in or supporting a practice acknowledged to cause extensive, unnecessary harm is, all else equal, wrong, even if the harm would occur without one's participatory actions or support. We would not excuse a member of a lynch mob on the grounds that his refraining from participating in the lynching would not be causally efficacious in reducing the harm to the lynched person. The Factory Harm and Environmental Harm arguments that appeal to principles (P3) similarly say that purchasing and consuming industrially-raised meat when vegetables are available is wrong, even if abstaining from such meat and eating vegetables instead would not be causally efficacious in reducing harm to animals or the environment from factory farming. In parallel, the roadkill arguments that appeal to the (P3) principles say that purchasing and consuming only vegetables when roadkill is available is wrong, even if abstaining from some of the vegetables and collecting and consuming roadkill instead would not be causally efficacious in reducing harm to animals or the environment from vegetable farming.

Opening the Floodgates

It might be admitted that ideally we should consume roadkill, but objected that doing so would lead us to desire to eat more meat than just the relatively

---

15. See the end of the section "Factory Farming and Roadkill" and note 9, where I address the amount of available roadkill.

16. See note 4 again for a little bit of detail.

small amount of roadkill available. Our psychological makeup (in particular, weakness of the will) would lead us to act on this desire, which would lead to greater harm to animals and the environment than refraining from eating meat altogether. So we should refrain from meat altogether.

This line of reasoning, which has been presented to me several times, is impressive for the number of reasoning fallacies it commits. First, it is a slippery slope argument. This argument has the very same bad form as an argument for celibacy that claims that we should not have sex with our committed relationship partners because doing so would stoke desires that would lead us, through weakness of the will, to having sex with many people, which is a worse condition overall than complete abstinence. Second, it shifts the burden of proof inappropriately, and in an especially objectionable way, by making an empirical claim about human behavior under certain conditions without any empirical evidence. The burden of proof should remain with the objector to show that certain widespread immoral actions would result from my proposal (though, indeed, statistically common monogamous partnerships would seem to be empirical evidence in favor of the assumption that we can avoid sliding down such slopes). Finally, the objection relies on a faulty dilemma. The objection assumes that if we eat meat, then either we eat roadkill or we eat immorally when we consume additional meat other than roadkill. But the Factory Harm and Environmental Harm Arguments at most establish that purchasing and consuming factory-farmed meat is immoral, not that consuming meat from non-factory farms or meat from hunted wild animals is immoral. Now in this chapter, I do not argue that these are morally acceptable sources of meat, but for an objector to infer on this basis that they are immoral sources of meat would be to engage the fallacy of negative proof.

Different or Additional Implications

Objection: Instead of showing what you say, it follows from the usual arguments for vegetarianism that we should put up fences on roadways to prevent animals from becoming roadkill.

Reply: That would likely be so costly that we would have to sacrifice many important human interests to pay for such fences. So I doubt it follows. Even if it does, surely it follows only in addition to, rather than instead of the claim that we should collect and consume roadkill, for we are presently in a situation without many fences but with much available roadkill. Not collecting and consuming it causes or supports a practice that causes extensive, unnecessary harm to animals and the environment, which is wrong by the standards of the harm principles in the arguments I have examined.

Objection: Okay, but your argument certainly shows that we should not drive our cars or fly in planes unless it is necessary, so it is immoral to drive to the movie theater or fly to philosophy conferences, for example. Otherwise we engage in a practice that causes extensive, unnecessary harm to animals and the environment. Similarly, your argument shows that we are obligated to engage in other freegan practices, such as collecting berries, foraging for unspoiled food in dumpsters, collecting and eating the dead bodies of field animals killed in vegetable production, as well as our dead pets and dead relatives. These are absurd implications of your argument.

Reply: Take care to notice that my thesis is not that we are obligated to collect and consume roadkill. My thesis is that the usual arguments for vegetarianism imply that we are obligated to collect and consume roadkill. Either the same reasoning does show the other things mentioned in the objection, or it does not. If it does not, then there is no problem for my thesis. If it does, then there are these further potential problems for the vegetarian who propounds the arguments I have examined, but still no problem for my thesis. So in neither case is there a problem for my thesis.

Health Risks

This objection takes issue with my claim above that road-killed meat is "perfectly nutritious." First, there is strong evidence that a pure vegetarian diet is nutritionally superior to diets containing meat, especially for the prevention of heart disease and many cancers. In reply, I would point out that, relative to most other meats, venison is very low in saturated fat and much higher in cholesterol-reducing polyunsaturated fat. As well, the research that is normally cited supporting the nutritional superiority of vegetarian diets does not address venison specifically.[17] Second, some deer carry Chronic Wasting Disease (CWD), the cervid version of Bovine Spongiform Encephalopathy (BSE, or mad cow disease). BSE is believed to cause the human variant, Creutzfeldt-Jakob Disease. In reply, I only state that although CWD is not fully understood, there are zero documented cases of humans contracting CWD and no scientific evidence that it is possible for humans to contract CWD. Third, one might worry about the risk of bacterial infection from

---

17. Note as well, in connection with the figures given in note 9, that under discussion is a relatively small amount of meat.

tainted roadkill. In reply, I reiterate that I have in mind only fresh, unspoiled, and intact wild animals.

Other Frequent Objections That I Can Treat Only Briefly

Objection: Roadkill is not harm free. Scavengers are deprived of food if we collect and consume it.

Reply: Usually large road-killed animals of the sort under discussion are not left to be scavenged, but are instead collected by government agencies and wasted in landfills. Individuals who collect roadkill for consumption can place the viscera and other parts not suitable for human consumption where they can be scavenged. Thus, far from harming scavengers, the collection of roadkill for consumption can benefit scavengers.

Objection: Collecting and consuming roadkill is disgusting.

Reply: Again, this is not a problem for my thesis. If it is a problem, it is a problem for the vegetarian whose arguments imply that we are obligated to collect and consume roadkill. I can offer some help in solving this problem, however. Even if it were true that collecting and consuming roadkill is disgusting, it would not affect any moral obligation we might have to collect and consume it. People are often morally obligated to do disgusting things. I once found myself obligated to clean my dog's disgusting vomit off of my grandmother's newly-installed carpet. Suppose you suffered a gruesome injury and a stranger could, with very little cost to herself, save your life by closing an impressively hemorrhaging gash in your flesh and applying pressure until the paramedics arrived. That might be disgusting, but (many claim) she would be obligated to do it regardless of the disgust. Some people feel disgust at the thought of eating non-animal sources of protein such as beans and tofu, yet we should not think this bears on whether they are obligated to refrain from animal protein. If one is obligated to do something disgusting, one should get over the disgust and fulfill one's obligation.

Conclusion

If the Factory Harm and Environmental Harm Arguments establish anything, it is only that we are obligated not to purchase and consume factory-farmed meat. Neither argument establishes that we are obligated to purchase and consume only vegetables. Drawing that conclusion requires the hidden premise that vegetables are the only morally acceptable alternative to factory-farmed meat. By the strict vegetarian's own standards, another alternative is roadkill.[18]

Not only is eating roadkill acceptable on the standards assumed by these arguments, it is obligatory, for otherwise one violates the harm principles central to these arguments. So not only do these arguments fail to establish what they are often alleged to establish, that vegetarianism is morally obligatory, but their failure is an interesting one because the core premises of those very arguments can be turned against them to support the conclusion that strict vegetarianism is immoral. The proponents of those arguments could escape this conclusion by giving up the harm principles, but then those arguments against eating factory-farmed meat would have to be given up as well, which is likely more than the vegetarian is willing to pay.[19]

References

Adams, Carol. 1990. *The Sexual Politics of Meat: A Feminist-Vegetarian Critical Theory.* New York: Continuum.

Blanco-Canqui, Humberto, C. J. Gantzer, S. H. Anderson, E. E. Alberts, and A. L. Thompson. 2004. "Grass Barrier and Vegetative Filter Strip Effectiveness in Reducing Runoff, Sediment, Nitrogen, and Phosphorus Loss." *Soil Science Society of America Journal*, 68(5): 1670–1678.

Bruckner, Donald W. 2007. "Considerations on the Morality of Meat Consumption: Hunted Game versus Farm-Raised Animals." *Journal of Social Philosophy*, 38(2): 311–330.

Curnutt, Jordan. 1997. "A New Argument for Vegetarianism." *Journal of Social Philosophy*, 28(3): 153–172.

Daniel, Carrie R., Amanda J. Cross, Corinna Koebnick, and Rashmi Sinha. 2011. "Trends in Meat Consumption in the USA." *Public Health Nutrition*, 14(4): 575–583.

---

18. Another alternative is insects. See C. D. Meyers (2013).

19. Earlier versions of this paper were presented at Chatham University, Penn State University, New Kensington, the Midsouth Philosophy Conference, the International Social Philosophy Conference, and a Group Session of the Society for Applied Philosophy. Thanks to Mylan Engel for his commentary at the Midsouth and to Robert Jones for his at the Society for Applied Philosophy. I have also benefited from feedback from and discussion with approximately one gazillion people, including most notably Maria Lasonen-Aarnio, Aaron Bell, Christopher Belshaw, Ben Bramble, Sean Bridgen, Lynne Dickson Bruckner, Mark Budolfson, Robert Farley, Bob Fischer, Alexa Forrester, Jennifer Gilley, Bart Gruzalski, Jeff Johnson, Alice Julier, Hanna Kim, Kerri LaCharite, Doug Lavin, Heather McNaugher, C. D. Meyers, Ben Minteer, Khrys Myrddin, Marc Nieson, Alastair Norcross, Howard Nye, Evan Riley, Tom Rumbaugh, Sally Scholz, Adam Stawski, Sheryl St. Germain, Joe Ulatowski, Jennifer Wood, Federico Zuolo, and anonymous reviewers for this volume.

Davis, Steven L. 2003. "The Least Harm Principle May Require That Humans Consume a Diet Containing Large Herbivores, Not a Vegan Diet." *Journal of Agricultural and Environmental Ethics*, 16(4): 387–394.

Dawkins, Richard. 2009. "Peter Singer—The Genius of Darwin: The Uncut Interviews" (interview). The Richard Dawkins Foundation for Reason and Science. Retrieved from http://richarddawkins.net/videos/3951-peter-singer-the-genius-of-darwin-the-uncut-interviews January 5, 2014.

DeGrazia, David. 2009. "Moral Vegetarianism from a Very Broad Basis." *Journal of Moral Philosophy*, 6(2): 143–165.

Denton, Andrew. 2004. "Professor Peter Singer" (interview transcript). *Enough Rope*. ABC News. Retrieved from http://www.abc.net.au/tv/enoughrope/transcripts/s1213309.htm January 5, 2014.

Diamond, Cora. 1978. "Eating Meat and Eating People." *Philosophy*, 53(206): 465–479.

Engel, Jr., Mylan. 2000. "The Immorality of Eating Meat." In Pojman, Louis P. (ed.), *The Moral Life: An Introductory Reader in Ethics and Literature*. New York: Oxford University Press. 856–890.

Engel, Jr., Mylan. 2011 [2005]. "Hunger, Duty, and Ecology: On What We Owe Starving Humans." In Pojman, Louis P., and Paul Pojman (eds.), *Environmental Ethics: Readings in Theory and Applications* (6th ed.). Belmont, CA: Wadsworth. 340–359. [Originally appeared in 4th edition, 2005.]

Forman, Richard T. T., Daniel Sperling, John A. Bissonette, Anthony P. Clevenger, Carol D. Cutshall, Virginia H. Dale, Lenore Fahrig, Robert France, Charles R. Goldman, Kevin Heanue, Julie A. Jones, Frederick J. Swanson, Thomas Turrentine, Thomas C. Winter. 2003. *Road Ecology: Science and Solutions*. Washington, DC: Island Press.

Fox, Michael Allen. 2000. "Vegetarianism and Planetary Health." *Ethics and the Environment*, 5(2): 163–174.

Iowa State University. 2009. "Beef and Pork Whole Animal Buying Guide." Small Meat Processors Working Group. Pamphlet.

Kendall, Gillian. 2011. "The Greater Good: Peter Singer on How to Live an Ethical Life." *The Sun Magazine: The Sun Interview*. May 2011, Issue 425.

Lamey, Andy. 2007. "Food Fight! Davis versus Regan on the Ethics of Eating Beef." *Journal of Social Philosophy*, 38(2): 331–348.

Meyers, C. D. 2013. "Why It Is Morally Good to Eat (Certain Kinds of) Meat: The Case for Entomophagy." *Southwest Philosophy Review*, 29(1): 119–126.

Neue, Heinz-Ulrich. 1993. "Methane Emissions from Rice Fields." *BioScience*, 43(7): 466–474.

Norcross, Alastair. 2004. "Puppies, Pigs, and People: Eating Meat and Marginal Cases." *Philosophical Perspectives*, 18(1): 229–245.

Pennsylvania Game Commission. 2014. "Pennsylvania White-tailed Deer: Deer Weight Chart." Retrieved from http://www.portal.state.pa.us/portal/server.pt/document/926461/deer_weight_estimating_chart_pdf January 5, 2014.

Rachels, Stuart. 2011. "Vegetarianism." In Frey, R.G., and Tom L. Beauchamp (eds.), *The Oxford Handbook of Animal Ethics*. Oxford: Oxford University Press. 877–905.

Rollin, Bernard E. 1995. *Farm Animal Welfare: Social, Bioethical, and Research Issues.* Ames: Iowa State University Press.

Singer, Peter, and Jim Mason. 2006. *The Way We Eat: Why Our Food Choices Matter.* Emmaus, PA: Rodale.

State Farm Insurance. 2011. "Deer/Elk/Moose Claims—Insurance Industry Projections 7/1/2002–6/30/2011." Unpublished report.

USDA Economic Research Service. 2014a. "Cattle & Beef / Statistics and Information." Retrieved from http://www.ers.usda.gov/topics/animal-products/cattle-beef/statistics-information.aspx January 5, 2014.

USDA Economic Research Service. 2014b. "Poultry & Eggs / Statistics and Information." Retrieved from http://www.ers.usda.gov/topics/animal-products/poultry-eggs/statistics-information.aspx January 5, 2014.

Vieux, Florent, Louis-Georges Soler, Djilali Touazi, and Nicole Darmon. 2013. "High Nutritional Quality Is Not Associated with Low Greenhouse Gas Emissions in Self-Selected Diets of French Adults." *The American Journal of Clinical Nutrition*, 97(3): 569–583.

Wang, Y., and M. A. Beydoun. 2009. "Meat Consumption Is Associated with Obesity and Central Obesity among US Adults." *International Journal of Obesity*, 33(6): 621–628.

3 THE ENVIRONMENTAL OMNIVORE'S DILEMMA

J. Baird Callicott

For most of the twentieth century, academic ethical theory has been confined within two complementary paradigms: utilitarianism and deontology, originating in the late eighteenth century in the work of Jeremy Bentham (1789) and Immanuel Kant (1785), respectively. Beginning in the 1970s, animal ethics were located in both paradigms by professional philosophers. The utilitarian version of animal ethics was pioneered by Peter Singer (1973, 1975); the deontological version by Tom Regan (1975, 1980, 1983). The former is called "animal liberation"; the latter "animal rights."

Environmental Ethics (the journal) began publishing in 1979. A paper of mine, "Elements of an Environmental Ethic: Moral Considerability and the Biotic Community," appeared in volume 1, number 1 (Callicott 1979). I had been teaching a course in environmental ethics at the University of Wisconsin–Stevens Point since 1971. Stevens Point is about ninety miles upstream from the location of Aldo Leopold's storied shack on the shores of the Wisconsin River—the place celebrated in his slender masterpiece, *A Sand County Almanac*. The climactic essay of that book is "The Land Ethic," and I had worked up a more systematic and philosophically explicit environmental ethic based on Leopold's amateur sketch.

Largely unbeknownst to academic philosophers, a biological paradigm of ethics had existed alongside the philosophical paradigms since the late nineteenth century, breaking out into notoriety with the publication of *Sociobiology: The New Synthesis* by Edward O. Wilson (1975). The biological paradigm is traceable to *The Descent of Man* by Charles Darwin (1871); and Darwin in turn had been largely informed by David Hume (1751), who had effectively set out his moral philosophy a quarter century before Bentham and Kant had done.[1] Thus three ethical paradigms had originated in the eighteenth century and all three thrived into the twentieth century, and now on into the twenty-first—two of them in the academic philosophical universe of discourse and the third in the biological. Naturally, given his background in the biological sciences, Leopold was informed by Darwin's biological paradigm of the origin and development of ethics in *The Descent of Man* and himself appears to be as innocent of the academic philosophical paradigms as his contemporaries in academic philosophy were of the biological.

The philosophical paradigms of ethics are militantly individualistic. In sharp contrast, a distinctive feature of the land ethic is its holism: "a land ethic," Leopold (1949: 204, emphasis added) wrote, "changes the role of *Homo sapiens* from conqueror of the land-community to plain member and citizen of it. It implies respect for his [individual] fellow members *and also respect for the community as such*." Indeed it is holistic with a vengeance because in the summary moral maxim of the land ethic, "fellow-members" drop out completely and it's all about "the community as such": "A thing is right when it tends to preserve the integrity, stability, and beauty of the biotic community. It is wrong when it tends otherwise" (Leopold 1949: 224–225).

And, back in the day, its holism was what made the land ethic the only *environmental* ethic then on offer because then environmental concerns were largely if not exclusively about transorganismic wholes—endangered *species*, ravaged *forests*, drained *wetlands*, polluted *rivers*, petroleum-fouled *seashores*, and so on. Environmentalists qua environmentalists then were not concerned about *individual* animals (and plants)—nor are they now, for that matter. As Leopold (1949: 107) bluntly put it, "The only certain truth," regarding the natural economy of the biotic community, "is that its creatures must suck hard, live fast, and die often, lest its losses exceed its gains." In "The Land Ethic" when Leopold (1949: 210–211) does say something about animal

---

1. I say "effectively" because Hume's *Treatise of Human Nature*, published in three volumes in 1739 and 1740, "fell stillborn from the press." And the Second Enquiry was a spruced-up version of Book III *On Morals*.

rights (a "biotic right . . . to continuance" for "songbirds . . . predatory mam-mals, raptorial birds, and fish-eating birds"), it is clear that rights devolve on them qua species, not qua individuals—maybe that's what makes such rights "biotic," I don't know. Leopold was an avid hunter to his last breath and his predilection for hunting was perfectly consistent with his land ethic, pro-vided that his hunting was both lawful and mindful (Flader 1974).

I was therefore distressed when I found that the new journal devoted to environmental ethics seemed to be publishing more papers about animal ethics than environmental ethics in the issues that followed the inaugural issue in which my paper had appeared.[2] So, I set out sharply to distinguish between animal ethics and environmental ethics with a now infamous screed titled "Animal Liberation: A Triangular Affair" (Callicott 1981). Under the influence of what might be called "Postmodern Primitivism"—beautifully set out by Paul Shepard (1973) in *The Tender Carnivore and the Sacred Game*—I wrote some things in that piece that were irresponsible. Perhaps most egre-gious was my declaration that the ideal human population would be twice the size of that of bears. (What kind of bears I did not stop to consider. All kinds I guess, although now it hardly matters what I was thinking back then.)

The retribution I deserved was soon forthcoming in the charge that the holistic Leopold land ethic, as I had theorized it, was a case of "ecofascism" or "environmental fascism" and would entail conclusions that only the most ex-treme and ruthless misanthropist could embrace. Tom Regan (1983) himself was the first to pummel me, but others—including William Aiken, Frederick Ferré, and Kristin Shrader-Frechette—soon piled on (see Callicott 1999 for citations and a full-tilt defense of the land ethic against the ecofascism charge). Look, it's simple and obvious: humans are "plain members and citizens" of the biotic community; hunting and killing deer, also plain members and citizens of the biotic community, which threaten its "integrity, stability, and beauty," when their populations are allowed to irrupt, is not just land-ethically permis-sible, it's land-ethically obligatory; grant for the sake of argument, that the root cause of our environmental problems is human overpopulation; so isn't it no less land-ethically permissible, nay obligatory, to cull the human popula-tion by whatever means necessary? And indeed, the land ethic *as I had theo-rized it* was guilty as charged.

But I had theorized the land ethic wrongly. My mistake turned on the misinterpretation of one word in the summary moral maxim—"when," which

2. In retrospect, I note that this is an exaggeration. The truth of the matter is that a few animal ethics papers appeared at the rate of about one per issue for the next four or five issues.

is not a logical operator. Again: "A thing is right *when* it preserves the integrity, stability, and beauty of the biotic community. It is wrong *when* it tends otherwise." The word "when" can be converted into a logical operator in either of two ways: it can be interpreted to mean "if" or it can be interpreted to mean "if and only if"—that is, it can be interpreted to specify a sufficient condition for a thing to be right or to specify both a sufficient and a necessary condition for a thing to be right. As I was pompously and self-righteously composing "A Triangular Affair," the difference had not occurred to me, but clearly I had interpreted "when" to mean "if and only if." But that was not what Leopold had meant by "when." If preserving the integrity, stability, and beauty of the biotic community is but a sufficient condition for a thing being right, and not also a necessary condition, then there may be other sufficient conditions for a thing to be right. A thing is right *if* it preserves the integrity, stability, and beauty of the biotic community; yes, but not also *only if* it preserves the integrity, stability, and beauty of the biotic community. A thing is also right if it does not violate human rights to life, liberty, and the pursuit of happiness.

What about the meaning of "when" in the second sentence? Same thing: a sufficient condition, not a necessary condition. A thing is wrong *if* it tends otherwise. But it is *not* wrong *if and only if* it tends otherwise. Other things can also be wrong for a different sufficient reason. For example, culling human beings by any means necessary would be wrong because it is a violation of the duties and obligations generated by our membership in the global village—the universal human community—from a communitarian point of view. Doesn't putting both sentences together—the first specifying a sufficient condition for a thing to be right, the second for a thing to be wrong—add up to a necessary and sufficient condition? No. One thing can be right if it preserves the integrity, stability, and beauty of the biotic community—say trimming the deer herd to the carrying capacity of the biotic community. Another thing can be wrong if it tends otherwise—failing to trim the deer herd. But trimming the human population by any means necessary can also be a sufficient condition for a thing to be wrong.

Another, less formal way of putting the point is this: The land ethic was not conceived by Leopold to be a replacement for all other modalities of human ethics; it was conceived to be an addition to them. After the land ethic is layered on, all other modalities of human ethics remain intact and in force. (And by "modalities of human ethics," I do not mean "either paradigms of ethics or the theories constructed within them." Mass murder is morally wrong, a thing contrary to human ethics. Ethical theories are offered up by philosophers—within the parameters of ethical paradigms, such as utilitarianism, communitarianism, and deontology—to explain why mass murder is wrong.)

How can we be so sure that Leopold meant "if" by "when," not "if and only if"? Because, following Darwin, Leopold conceives of ethics developing sequentially. In Leopold's "ethical sequence," he thinks—wrongly in my opinion (for reasons that I get to shortly)—that "The first ethics dealt with the relation between individuals" while "[l]ater *accretions* dealt with the relation between the individual and society" (Leopold 1949: 202–203, emphasis added). Leopold conceives the land ethic to be the next step in this series of "accretions." An accretion is a layering over, a laminate. A good example of a series of accretions would be tree rings, each laid down over the previous one, annually extending the girth of the tree trunk. Ethics does not expand like a balloon, leaving no trace of its previous boundaries, as in Singer's (1981) *Expanding Circle*, but rather like the way a tree expands.

Darwin (1874: 123), contrary to Leopold, thought—correctly in my opinion—that "actions are regarded by savages and were probably so regarded by primeval man, as good or bad, solely as they obviously affect the welfare of the tribe—not that of the species, nor that of an individual member of the tribe." As we see, Darwin regards holistic ethical sensibilities to be more basic than individualistic ethical sensibilities and that "ethics [which] dealt with the relation between individuals" were a later accretion. The individualistic and holistic aspects of a tribal ethic are, of course, hard to pull apart. Acts of "murder, robbery, treachery, etc." are "crimes" perpetrated by one individual on others (Darwin 1874: 120). Such acts would, however, have an adverse effect on the welfare of the tribe, as well as on the welfare of the individuals who suffer from them, for, as Darwin (1874: 120) observes, "No tribe could hold together," if such crimes "were common." The institution of human sacrifice, prevalent among many peoples in the past, suggests that the welfare of the tribe was not only paramount, but the only consideration, just as Darwin supposes.

Darwin goes on to suggest the following "ethical sequence." Ethics originates, he thinks—plausibly in my opinion—when the mammalian "parental and filial affections" spill beyond parent and offspring to grandparents, siblings, uncles, aunts, and cousins, enabling an extended family or clan to incorporate and pursue life's struggle collectively (Darwin 1874: 109). Competition among clans for limited resources made it advantageous in the struggle for survival and reproductive success for clans to merge and form larger and therefore more powerful social wholes—tribes, in a word. And with the emergence of tribes, the moral sentiments would be extended to the larger social whole and to its members, "but excite no such sentiment[s] beyond these limits" (Darwin 1874: 120). This selective pressure to merge smaller societies into larger and more complex societies—with a corresponding extension of ethics—continued to operate until, by Darwin's time, in Europe and Greater Europe (the Americas and Australia) nation states had evolved, with such peculiar and correlative moral sentiments as patriotism. Darwin himself envisioned this process to reach a climax with the inclusion of all humanity in the purview of ethics:

> As man advances in civilization, and small tribes are united into larger communities, the simplest reason would tell each individual that he ought to extend his social instincts and sympathies to all the members of the same nation, though personally unknown to him. This point being once reached, there is only an artificial barrier to prevent his sympathies extending to the men of all nations and races (Darwin 1874: 126–127).

Ethics evolved in other ways as well. If ethics originates as a means of holding groups together, then the content of ethics—just what is right and what is wrong—changes in response to changes in the lifeways and economies of those groups, as well as to changes in the size of their membership. It would stand to reason, therefore, that a strong emphasis on private property rights would emerge only when the economy of a group shifted from a foraging to an agrarian lifeway (Mauss 1990). Also—in the order opposite to the one Leopold thought had happened—ethics, in Darwin's account, became more individualistic (as the words just quoted amply indicate).

While Leopold's notions of the origin of ethics and the ethical sequence (despite his getting it upside down) are doubtless due to the influence of Darwin on his thinking, based on the following passage in the Second Enquiry, we can also be equally confident that Darwin's notion of the ethical sequence is due to the influence of Hume on his thinking:

> But suppose the conjunction of the sexes be established in nature, a family society immediately arises, and particular rules being requisite for its subsistence, these are immediately embraced, though without comprehending the rest of mankind within their prescriptions. Suppose that several families unite together into one society which is totally disjoined from all others, the rules which preserve the peace and

order enlarge themselves to the utmost of that society, but . . . lose their force when carried one step farther. But again suppose that several distinct societies maintain a kind of intercourse for mutual convenience and advantage, the boundaries of justice grow still larger, in proportion to the largeness of men's views and the force of their mutual connections. History, experience, reason sufficiently instruct us in this natural progress of the human sentiments. (Hume 1751: 23)

Now, back to the problem of ecofascism and the land ethic. Our social memberships remain plural and hierarchically structured. Nuclear families belong to extended families. Let's say the Sean-and-Sinead-Mulligan family of four, living in Crescent City, State of Jefferson, belongs to the Mulligan clan scattered across North America, the British Isles, and Australia. The Mulligan clan belongs to the Irish nation, but the clan members are now citizens of several nation states—say, Ireland, the United Kingdom, the United States, Canada, and Australia. And all the Mulligans are denizens of the Global Village. On top of this hierarchy, Aldo Leopold has convinced the Crescent-City Mulligans that they are plain members and citizens of the Klamath River Watershed biotic community, nested in the larger Pacific Northwest temperate rainforest biome, which is, in turn, nested in the global biosphere.

So: out of the frying pan of ecofascism and into the fire of a bewildering kind of moral pluralism. All of these community memberships simultaneously generate duties and obligations. And to fulfill the duties and obligations generated by one community membership might mean neglecting—or worse, violating—the duties and obligations generated by another community membership. What are Sean and Sinead Mulligan to do if they want to do what's right?—given that lots of things they might do are right, but they cannot do all of them. Who should the Crescent-City Mulligans root for if Ireland and the United States play one another in a World Cup football match? More seriously, to which side of the conflict between spotted-owl preservation and the jobs of mill workers should the Crescent-City Mulligans throw their support? Should Sean and Sinead lavish extravagant Christmas presents on their children, Shannon and Seamus, or give them fewer and more modest presents at Christmas and donate the money saved to Oxfam?

I offered a protocol for resolving such conflicts (Callicott 1999). In reflecting on my own moral sensibilities in such di-, tri-, and quatra-lemmas and on my observations of what others do and say, it seems to me that the more intimate and venerable of our community memberships are the more insistent. For example, if one's duty to stay home and care for a sick relative is in

conflict with one's obligation to go to a city council meeting to protest subur-
ban hydraulic fracturing, the family duty would seem to take precedence over
the civic duty. We also weigh and factor into our moral deliberations, in such
circumstances, what might be called strength-of-interest. One has a duty to
provide joy as well as sustenance for one's children, but one's children's interest in a trip to Disneyland is weaker than the interests of one's destitute fellow
citizens in food and shelter; and so one's duty to pay one's taxes to support
social safety nets precludes one from avoiding paying one's taxes and spending
the savings on a family vacation to Disneyland. In general, first prioritize
duties and obligations generated by the more intimate and venerable community memberships in cases of conflicting duties and obligations, but then
prioritize the duties and obligations of the more impersonal and recently
formed community memberships when their stronger interests are at stake.
For example, if one's nation state confronts what is now known as an "existen-
tial threat," then one might well leave one's family and volunteer to serve in
one's country's armed forces.

So how do *domestic* animals fit into this communitarian moral landscape
(assuming that Leopold has adequately fit wild animals into it)? Mary Midgely
(1984) provides the answer: Domestic animals are members with us of "mixed
communities." As Midgely (1984: 112) sagely points out,

> All creatures which have been successfully domesticated are ones
> which are originally social. They have transferred to human beings the
> trust and docility which, in a wild state, they would have developed
> towards their parents, and in adult life towards the leader of their
> pack. . . . They became tame, not through the fear of violence, but be-
> cause they were able to form individual bonds with those who tamed
> them by coming to understand the social signals addressed to them. . . .
> They were able to do this, not only because the people taming them
> were social beings, but because they themselves were so as well.

Thus, making moral matters more complicated still, many of us are members of multiple mixed communities. I live in Denton County, Texas, which
has the largest horse population of any county in the state (Backstrom 2002).
Many Denton-County ranches also harbor cattle and almost all of them are
home to dogs. The typical mixed human-dog community in Denton County
is almost familial—dogs are treated here, as they are in most of the United
States, like second-class children, living literally cheek by jowl with actual
family members. Beef is a staple of the Texas diet and many ranchers raise

their own "grass-finished" beef without the use of hormones or antibiotics.[3] But possessing horsemeat with the intent to sell it as food for human consumption—even in overseas markets—is a crime in Texas.[4] The legal difference between horsemeat and beef has nothing to do with human health; rather, it has everything to do with the difference in the mixed communities constituted by cattle and horses. Horses are not, like dogs, primarily "companion animals" any more than they are primarily meat animals. They are associates in rural work and play—they once pulled plows, wagons, and carriages and "cut" cattle (some still do); they play polo, prance around in horse shows, race, jump obstacles, and, more problematically, play various roles in rodeos.

By contrast with this nuanced Humean-Darwinian-Leopoldian-Midgelyan biological paradigm of ethics, animal liberation and animal rights are ham-handed (pun intended). They proceed on a quaint mode of thinking analogous to essence-accident metaphysics in ancient and medieval moral philosophy—a mode of thinking that will surely be abandoned as twenty-first-century moral philosophy matures. An essential property—*qua moral consideration* not qua metaphysical essence—is identified and justified. In the case of classical Singerian animal liberation that essential property is sentience; in the case of classical Reganic animal rights, it's being the subject-of-a-life. All other properties are—*qua moral consideration*—accidents: gender, race, IQ, age, species. All beings that possess the morally essential property equally deserve equal consideration of their interests. One size fits all. Morally speaking, humans = cattle = deer = rats = ...

So what follows from these three paradigms of animal ethics—utilitarian, deontological, and communitarian—when the question is What to eat and what not to eat? (For simplicity's sake, I leave out of account virtue ethics, which has become ever more prominent in moral philosophy since the publication of *After Virtue*, by Alasdair MacIntyre in 1981, but has had a more modest impact on animal ethics.)

Let's start with the communitarian paradigm (not forgetting that it is pluralistic at the level of application). Omnivory is consistent with Midgleyan mixed-community animal ethics, provided certain important constraints are observed. The genesis of domestic animals is not, as I once supposed, a matter

---

3. All cattle are "grass fed," but most are shipped to feedlots for "finishing" on a diet of grains, which "marbles" their muscle tissue with fat.

4. http://www.statutes.legis.state.tx.us/Docs/AG/htm/AG.149.htm. And see *Empacadora de Carnes de Fresnillo, S.A. de C.V. v. Curry*, 476 F.3d 326 (5th Cir. 2007).

of involuntary capture, enslavement, and then selective breeding, but more a matter, as Midgely (1984) suggests, of an implicit social contract (Clutton-Brock 1999)—what Stephen Budiansky (1992) calls a "covenant of the wild." It is easy to imagine a bold and cunning wolf-pack leader establishing a mutualistic relationship with human hunters pursuing the same prey as do he and his subordinates, the ancestors of the domestic dog. But what possible "covenant" could animals form with human beings who would eventually slaughter and eat them?—how could any such asymmetrical relationship be mutually beneficial? Humans could protect such animals as the wild ancestors of domestic cattle and swine from predation, shelter them from the elements, and feed them when otherwise they might starve. The bargain, from the animal's point of view, would be a better life at the price of a shorter life. Denton County ranchers are generally affluent (often from oil money) and run small-scale as well as up-scale operations. One might say that they are in business for their health. And they faithfully keep the ancient covenant with their livestock. The noble horses are pampered, meticulously exercised, trained, and groomed; the cattle are well cared for during their short lives, transported humanely to a local processor, and there slaughtered painlessly. Provided one eats beef and the flesh of other domestic animals with whom the social contract remains operative, the communitarian paradigm of animal ethics permits omnivory. Eating beef and other meat produced in factory farms is impermissible, from the perspective of communitarian domestic-animal ethics, because the social contract of the mixed community has been abrogated. What we might call "covenant viands" are more expensive—often much more expensive—but sometimes a good conscience comes at premium price.

Omnivory would seem to be inconsistent with most deontological animal ethics as Palmer (2010) indicates. Certainly its chief and iconic exponent, Tom Regan (1983), thinks that it is, because foremost among the rights that he attributes to animals is the right to life. That right, however, leads to untoward environmental consequences. For we humans, presumably, would have a duty to protect rights-owning herbivorous wild animals from their carnivorous wild predators.

Countering this putative fatal environmental flaw in his deontological animal ethics, Regan (2013) argues that predatory animals are not moral agents and thus can do no wrong when they attack and kill other animals. But that misses the point. Consider the following analogy. Some human beings kill other human beings. A small subset of such killers are clinically insane. If insane killers' attorneys successfully mount the "insanity defense," they are not *punished* by imprisonment or death, as are convicted murderers, precisely because they are judged not to be responsible for their actions, not to be moral agents. But neither are they freed and released into the human community to go on killing. Rather they are detained in secured institutions to prevent them from continuing to violate the right to life of other human beings. Thus if animals (or, more precisely, mammals, in Regan's theory) have a right to life, then carnivorous wild animals should be prevented from killing innocent wild mammalian herbivores in the biotic community. Like clinically insane killers, carnivores (such as felines) and omnivores (such as badgers) should be locked up in secure institutions (such as zoos), prevented from reproducing, fed artificial foods made of vegetable products, and allowed to live thus until they die without issue. That of course would wreak ecological havoc and be tantamount to the deliberate extinction of some of the most charismatic of extant species.

Regan (2013) attempts to evade the inexorable logic of this analogy by another. Consider a mountain lion attacking a mule deer and then consider a mountain lion attacking a human *child*. We do, Regan (2013) says, have a duty to protect the child, but not the deer, from being killed by the lion. That's because the deer is "competent" to secure its own right to life and that of its young; so we would patronize it were we to protect it and its young from lion predation, while the child is not competent to secure its own right to life and we thus have a duty to protect it. But protection of humans from mountain-lion predation is by no means confined to children. We do everything we can to prevent mountain lion attacks on humans, adults as well as children (Subramanian 2009). The occasional lion that kills an adult human is hunted down, if possible, and shot dead (Deubrouk 2007). If deer rights = human rights—and that's the theory of animal rights—don't deer deserve the same protection from predation by wild animals as humans?

Surprising to some perhaps, omnivory would be not just consistent with utilitarian animal ethics, but a positive duty. If humans collectively and universally decided to quit eating meat, the populations of domestic animals that are raised for human consumption would precipitously plummet and bottom out at a few-thousand museum specimens. Why? They presently exist in such numbers—1.5 billion cattle, one billion swine, one billion sheep—for but one reason: because they are produced for human consumption.[5] (Please bear with me; I'm getting to the positive duty to eat meat from the point of view of utilitarian animal ethics.) Most animals raised for human consumption are

---

5. The Economist Online, July 27, 2011 (accessed March 6, 2014) http://www.economist.com/blogs/dailychart/2011/07/global-livestock-counts.

not treated like the cattle on Denton County ranches are treated. Let me stipulate that factory-farmed animals endure a life in which the suffering they experience exceeds the pleasure they experience and thus that, from a sentience-based utilitarian point of view, they would be better off not to exist than to exist. From a preference-based utilitarian point of view, the case would be the same: factory-farmed animals are seldom able to satisfy their preferences; and they experience a life that is, on balance, one of acute preference frustration. So, for humans collectively and universally to decide to quit eating meat would be the right thing to do, according to utilitarian animal ethics. But that humans will collectively and universally decide to quit eating meat is less likely than that we will collectively and universally decide to eliminate the abuses of factory farming and require the implementation of humane methods of animal agriculture (Chambers and Grandin 2004). In which case, the pleasure and/or preference satisfaction that livestock experience might well exceed the pain and/or preference frustration they suffer. And, if so, thus to maximize utility—the pleasant/preference-satisfaction-fulfilled short life that meat animals enjoy + the many pleasant/preference-satisfaction-fulfilled experiences of meat-eating long-lived humans—those persuaded by utilitarian animal ethics should become enthusiastic omnivores. Even without elimination of factory farming, those persuaded by utilitarian animal ethics can, as things now stand, eat meat in good conscience provided, like I, that they can assure themselves that the animals they are eating were raised in pleasant circumstances and were humanely transported and slaughtered. Indeed, they have a duty to eat humanely produced meat, thus doing their bit to maximize utility.

Also surprising to some perhaps, omnivory is even more severely constrained by the land ethic than by the mixed-community Humean-Darwinian-Midgleyan animal ethic. Why? Because animal agriculture, even with the very best practices, is necessarily inefficient by an order of magnitude because herbivorous animals convert only about 10 percent of the plant matter they metabolize into their own body parts (Pimentel and Pimentel 2007). On top of that, many of their body parts are not eaten by humans. If people ate only vegetable foods and the flesh of wild animals, "harvested" under strict regulations to ensure optimum sustained yield, lands now devoted to growing grain and pasturing domestic livestock could be retired and restored to a wild condition. The biotic community could expand tenfold, and the type of mixed community composed of humans and their consumable livestock would shrink to museum-sized patches here and there.

I have also recently worked out a new Earth ethic in response to the moral challenge of global climate change that would reinforce the constraints on omnivory that the land ethic would impose (Callicott 2013). The principal cause of global climate change is an increasing amount of carbon dioxide in the atmosphere (IPCC 2013). The metabolisms of livestock produce a significant amount of carbon dioxide and also a significant amount of methane, an even stronger, but more ephemeral, greenhouse gas (IPCC 2013). But of course so do the metabolisms of wild animals, especially ungulates. I am unable to do the calculations reliably to estimate the net reduction in greenhouse gases that a worldwide reduction of the livestock population to museum-sized remnants would yield, but I think it would be significant because domestic animals exist in greater densities than do wild animals—and thus there would be many fewer animals in the world in that scenario.

Scenarios, however, are just what the foregoing speculations set out—and improbable ones at that. We philosophers are free to indulge in "thought experiments"—what would happen if: animal rights were universally acknowledged and strictly enforced; certifiably humane animal agriculture were to become a universal practice; all people everywhere started to take climate change seriously and implemented an Earth ethic? But we, philosophers and non-philosophers alike, live in the real world, and many of us want to do what's right—to be ethical persons, moral human beings—as things now stand. Implicit in the part of this chapter focused on the problem of ecofascism for the land ethic, there is a general theory of ethics, an overarching moral-philosophy paradigm at play here that I call "communitarian." The duties and obligations that most people palpably feel and to which they usually actually respond are generated by our multiple community memberships. A few philosophers and those in their thrall may try to put single-principle utilitarianism (maximize aggregate utility) or deontology (treat others as ends withal and never as means only—"others" including a specified subset of animals) into practice, but doing so leads to all sorts of maladroit actions. What would actually happen if one's every action were determined by a strict adherence to Bentham's hedonic calculus or to Mill's greatest happiness principle? What would actually happen if one's every action were determined by the categorical imperative in defiance of one's "inclinations?"

In the context of what's right to eat, let me tell you what I observed to actually happen, and you be the judge.

I recently went to lunch with several academic philosophers, one of whom was an ethical vegan. There were vegetarian options on the menu and he chose one, but then asked the server if the entrée were not only vegetarian but also vegan. The server did not know. She was sent back to the kitchen to inquire of the chef. And while she was off on this errand, the vegan expressed indignant irritation at this poor minimum-wage worker's ignorance, complaining that she should know all the ingredients and their provenance of all the items on the menu. A duty to the ill-applied categorical imperative or utilitarian *summum bonum* (I did not ask which) resulted in a violation of a duty to colleagues (who practice communitarianism, some despite their theoretical loyalties, and who were embarrassed by the vegan's churlishness) and a duty to be kind to a person in a subordinate position during a social interaction. As a gesture of compensation for the inconsiderate behavior of the vegan, I left a bigger-than-usual tip on the table.

Here's another anecdote. I recently attended a dinner party at the home of a very senior member of another department and his wife, along with a few other faculty and a few of their graduate students. Our hosts were not themselves vegans or even vegetarians, but they had, in consideration of those who might be, prepared a vegan meal. Two of the graduate students declined to dine, nevertheless, and sat stolidly in front of empty plates. They explained that they had eaten before they came, not knowing if the food they would be served met their ethical standards.

Communitarianism is paradigmatically monistic (duties and obligations are generated by community membership) and practically pluralistic (we are simultaneously members of multiple communities—familial, municipal, national, global, mixed, biotic—and so are importuned by multiple and often conflicting duties and obligations, which we are obliged to prioritize for purposes of coherent moral action). It originates with the moral philosophy of David Hume (1751: 92), who excoriates "[c]elibacy, fasting, penance, mortification, self-denial, humility, silence, solitude, and the whole train of monkish virtues." Communitarinism was naturalized by Charles Darwin and applied to the biotic community by Aldo Leopold and to the mixed human-animal communities by Mary Midgley. Presently it has been vindicated by evolutionary moral psychology (Boehm 2012, Bowles and Gintis 2011, Haidt 2007).

Especially from the point of view of environmental ethics—both the land ethic and the Earth ethic—to eliminate animal agriculture and to reduce the numbers of domestic livestock to remnant populations maintained as historical curiosities would be the right thing for us *collectively* to do. And, should that ever come to pass, while humans might land-ethically and Earth-ethically remain omnivores, venison and other wild viands would be a rare treat (except for the very wealthy), obtained at a high price. Whether or not, by some miracle, that comes to pass, it remains true that we humans are members of many other communities than the biotic and biospheric communities. And they too engender duties and obligations, which must be weighed on the same scale with the duties and obligations engendered by our biotic and biospheric community memberships.

Personal and voluntary abstinence from the products of animal agriculture is as ineffectual a response to the alarming conversion of wildland to cropland as it is to the ever-increasing emission of carbon dioxide and methane into the atmosphere (Sinnot-Armstrong 2010). At planetary scales, the impact of one's personal and voluntary abstinence from meat is negligible. The conversion of wildland to cropland and the accumulation of greenhouse gases in the atmosphere are environmental problems created by the course of human bio-cultural evolution and can be effectively addressed only by *collective* human action. There may be some symbolic value in being a strict herbivore, but, hewing to one's vegan virtue at the cost of violating one's social duties and obligations generated by various human-community memberships may be counterproductive. Certainly it proved to be counterproductive in the anecdotes I just related. Had it not been inappropriate to do so, I would like to have invited that vegan-abused server out for a (grass-finished, organic) steak dinner; and had it not been inappropriate to do so, I would like to have insisted that those two elder-insulting graduate students take my seminar in ethical theory, thus to learn that there are more ways to think about ethics than what is possible to think within the constraints of the two prevailing paradigms—utilitarianism and deontology—to which moral philosophy has been largely limited of late.

References

Backstrom, Gayle. 2002. *I'd Rather Be Working* (New York: American Management Association).

Bentham, Jeremy. 1789. *An Introduction to the Principles of Morals and Legislation* (Oxford: The Clarendon Press).

Boehm, Christopher. 2012. *Moral Origins: The Evolution of Virtue, Altruism, and Shame* (New York: Basic Books).

Bowles, Samuel, and Herbert Gintis. 2011. *A Cooperative Species: Human Reciprocity and Its Evolution* (Princeton, N.J.: Princeton University Press).

Budiansky, Stephen. 1992. *Covenant of the Wild: Why Animals Chose Domestication* (New York: William Morrow).

Callicott, J. Baird. 1979. "Elements of an Environmental Ethic: Moral Considerability and the Biotic Community," *Environmental Ethics* 1: 71–81.

Callicott, J. Baird. 1981. "Animal Liberation: A Triangular Affair," *Environmental Ethics* 2: 311–338.

Callicott, J. Baird. 1999. "Holistic Environmental Ethics and the Problem of Ecofascism," in J. Baird Callicott, *Beyond the Land Ethic: More Essays in Environmental Philosophy* (Albany: State University of New York Press). 59–76.

Callicott, J. Baird. 2013. *Thinking Like a Planet: The Land Ethic and the Earth Ethic* (New York: Oxford University Press).

Chambers, Philip G., and Temple Grandin. 2004. *Guidelines for Humane Handling, Transport and Slaughtering of Livestock* (Bangkok: Food and Agriculture Organization of the United Nations).

Clutton-Brock, Juliet. 1999. *A Natural History of Domesticated Mammals*, Second Edition (Cambridge: Cambridge University Press).

Darwin, Charles R. 1871 First Edition/1874 Second Edition. *The Descent of Man and Selection in Relation to Sex* (London: John Murray). Page numbers refer to Charles Darwin, *The Descent of Man*, Second Edition (Amherst, N.Y.: Prometheus Books, 1998).

Deurbrouk, Jo. 2007. *Stalked by a Mountain Lion: Fear, Fact and the Uncertain Future of Cougars in America* (Guilford, Conn.: Globe Pequot).

Flader, Susan L. 1974. *Thinking Like a Mountain: Aldo Leopold and the Evolution of an Ecological Attitude toward Deer, Wolves, and Forests* (Madison: University of Wisconsin Press).

Haidt, Jonathan. 2007. "The New Synthesis in Moral Psychology," *Science* 316: 998–1002.

Hume, David. 1751. *An Enquiry Concerning the Principles of Morals* (London: A Millar). Page numbers refer to David Hume, *An Inquiry Concerning the Principles of Morals*, Charles W. Hendel, ed. (New York: Library of the Liberal Arts, 1957).

IPCC. 2013. *Climate Change 2013: The Physical Science Basis. Contribution of Working Group I to the Fifth Assessment Report of the Intergovernmental Panel on Climate Change* (New York: Cambridge University Press).

Kant, 1785. *Grundlegung zur Metaphysik der Zitten* (Berlin: L. Heimann).

Leopold, 1949. *A Sand County Almanac and Sketches Here and There* (New York: Oxford University Press).

Mauss, Marcel. 1990. *The Gift: The Form and Reason for Exchange in Archaic Societies* (London: Routledge).

Midgely, Mary. 1984. *Animals and Why They Matter* (Athens: University of Georgia Press).

Palmer, Clare. 2010. *Animal Ethics in Context* (New York: Columbia University Press).

Pimentel, Marcia H., and Pimentel, David. 2007. *Food, Energy, and Society*, Third Edition (New York: CRC Press).

Regan, Tom. 1975. "The Moral Basis of Vegetarianism," *Canadian Journal of Philosophy* 5: 181–214.

Regan, Tom. 1980. "Animal Rights, Human Wrongs," *Environmental Ethics* 2: 99–120.

Regan, Tom. 1983. *The Case for Animal Rights* (Berkeley: University of California Press).

Regan, Tom. 2013. "Animal Rights and Environmental Ethics," in Donato Bergandi, ed., *The Structural Links between Ecology, Evolution, and Ethics: The Virtuous Epistemic Circle* (Dordrecht: Springer). 117–126.

Shepard, Paul. 1973. *The Tender Carnivore and the Sacred Game* (New York: Charles Scribner's Sons).

Singer, Peter. 1973. Review of *Animals, Men, and Morals*, edited by Stanley Godlovitch, Roslind Godlovitch, and John Harris, published in the *New York Review of Books* (April 5).

Singer, Peter. 1975. *Animal Liberation: A New Ethics for Our Treatment of Animals* (New York: The New York Review).

Singer, Peter. 1981. *The Expanding Circle: Ethics, Evolution, and Moral Progress* (New York: Farrar, Straus, and Giroux).

Sinnott-Armstrong, Walter. 2010. "It's Not My Fault: Global Warming and Individual Moral Obligations," in Stephen Gardiner, Simon Caney, Dale Jamieson, and Henry Shue, eds., *Climate Ethics* (New York: Oxford University Press): 332–346.

Subramanian, Sushma. 2009. "Should You Run or Freeze When You See a Mountain Lion?" *Scientific American*, April 14 http://www.scientificamerican.com/article/should-you-run-or-freeze-when-you-see-a-mountain-lion/ (accessed March 8, 2014).

Wilson, Edward O. 1975. *Sociobiology: The New Synthesis* (Cambridge, Mass., The Belknap Press of Harvard University).

CHALLENGING MEAT

4 INDIVIDUAL CONSUMPTION AND MORAL COMPLICITY

Julia Driver

Consider the following case:*

> Aubrey believes that sentient animals have moral standing and
> that, if eating meat harms animals, one ought not to eat meat.
> However, she has recently read articles that indicate that her *in-
> dividual* choice to eat meat on *any given occasion* makes no
> difference—it in no way reduces animal suffering. However, she
> also believes that if she were to eat meat, she would be *part of*
> something that does lead to devastating harms to animals. To
> her, the thought of eating meat is also reprehensible because of
> this—because it involves the participation in a *collective* wrong-
> doing. Even if her individual act makes no difference to any par-
> ticular animal, it seems somehow "wrong" to her because it em-
> bodies the wrong sort of attitude toward such an evil. One
> ought to stand up for the good and against the bad, and buying
> into a bad collective practice seems utterly opposed to this ideal.

* Some of the material in this chapter was presented in "Moral Complicity," the 3rd Annual
Gertrude Bussey Lecture at Northwestern University, March 5, 2014. I thank the audience
members for their very helpful comments. This material was also discussed in two blog posts:
at PEA Soup, October 21, 2013, and at Political Philosop-her, February 7, 2014. I would also
like to thank Bob Fischer for his written comments on an earlier draft of this chapter.

This, in a nutshell, captures the dilemma of someone—consequentialist or not—who professes to accept what Christopher Kutz terms the Individual Difference Principle (IDP):

> (Basis) I am accountable for a harm only if what I have done made a difference to that harm's occurrence. (Object) I am accountable only for the difference my action alone makes to the resulting state of affairs.[1]

Compare Aubrey's case with the case of Blake:

> Blake believes that sentient animals have moral standing and that, if eating meat harms animals, one ought not to eat meat. However, she has recently read articles that indicate that her *individual* choice to eat meat on *any given occasion* makes no difference—it in no way reduces animal suffering. She also believes that if she were to eat meat, she would be *part of* something that does lead to devastating harms to animals, however, since her individual choice makes no difference, she decides to eat meat. Also, precisely because she does care about reducing animal suffering she makes sure to purchase meat *only* from large industrial factory farms, rather than small, local, more humane producers. She does this because she believes that her decision to buy meat from a small producer is more likely to have an impact on production practices. She believes that she will be the cause of suffering if she buys more humanely produced meat. Therefore, she should not buy humanely produced meat, but, rather, buy meat produced on a massive industrial scale in which huge suffering is the norm. But, as long as she causes none of it, that is perfectly okay.

I think that most people will find Audrey more sympathetic than Blake, and yet Blake's reasoning reflects the reasoning of many who would argue that even though suffering is bad, eating meat in the modern world is not wrong because when an individual buys a specific piece of meat that individual is not making a purchase that impacts production practices. Thus, due to IDP, no harm, no foul.

In this chapter I will try to give an account of why Blake is wrong, an account that bypasses IDP. While the account is compatible with certain forms of *global* consequentialism, it is not committed to it. Global consequentialism holds that more than just actions are evaluable according to the

---

1. *Complicity: Ethics and Law for a Collective Age* (Cambridge University Press, 2000), 116.

consequentialist standard. The version that I favor also regards states relevant to agency as so evaluable, including character traits. Thus, a virtue is a character trait that produces good systematically.[2] This gives the theory more evaluative flexibility.

Writers like Kutz view IDP as a commitment of consequentialism. This is not true—there are many writers sympathetic to consequentialism who reject it. Further, non-consequentialists will often find IDP appealing—it seems, if anything, a feature of common-sense morality, as Kutz himself points out.

Some writers hold that even if, for the sake of argument, we grant that eating meat is morally problematic (in some sense), when an individual eats a hamburger (for example), that *individual* has done nothing wrong, at least on straightforward *consequentialist* grounds. The argument tends to take two forms: the simple no-difference view, and the somewhat more sophisticated, *no-trigger* variant of the no-difference view.[3] In this chapter I will assume that these arguments work insofar as they show that there are circumstances in which one person deciding to eat meat *on a given occasion* makes no difference to production policies (and no expected difference, and does not raise the expected disutility in any way either straightforwardly or via a triggering mechanism), and thus no difference to the total amount of animal suffering. However, even in light of this, I will argue that we have grounds to hold Blake wrongfully complicit in the harms of meat production.

Moral Complicity and Causal Impotence

In *Complicity*, Christopher Kutz argues that consequentialists have trouble with cases of wrongful complicity in which the complicit act makes no relevant difference to the outcome in question. This is because of their commitment to IDP.

In spelling out IDP, Kutz distinguishes the Basis from the Object in the following way: "By the basis of accountability, I mean the facts about agents that warrant holding them accountable for a harm or a wrong."[4] The object of accountability is different—two accountability acts can have the same basis

---

2. See my *Uneasy Virtue* (Oxford University Press, 2001).

3. Julia Nefsky has recently argued that Shelly Kagan's "trigger" argument fails. See her "Consequentialism and Collective Harm: A Reply to Kagan," *Philosophy and Public Affairs* 39 (2012), 364–395. Kagan's argument appears in "Do I Make a Difference?," *Philosophy and Public Affairs* 39 (2011), 105–141.

4. Kutz, 115.

but differ in terms of object. He asks us to compare the case of an agent hitting someone, versus the same agent telling his brother to hit someone. In each case, the basis is the same—it is still a *fact* about *that agent*, and not someone else. However, the objects are different since in one case it is the agent's hitting, and in the other case it is the brother's hitting that are the objects of accountability.[5] This is an important distinction to Kutz because it accounts for some confused claims about accountability—he thinks that while basis claims need to be individualistic, object claims need not be. Thus, it can be that a fact about *me* (that I ate the hamburger) that makes *me* accountable in the individualistic sense, but in the object sense it may not be my specific act that is the focus of accountability; rather, it can be a collectively produced harm. IDP is problematic, on his view, because both are understood individualistically, which makes it an unsuitable principle for understanding collective wrongdoing and individual accountability within collective wrongdoing.

As mentioned earlier, IDP has been contested, and has been contested by writers at least sympathetic to consequentialism, so it is a mistake to hold that consequentialists are *committed* to it.[6] Further, very many non-consequentialists find something like IDP intuitively plausible—how can I be held responsible for a bad outcome if I haven't done anything to cause that outcome?

Kutz critiques IDP by discussing causal overdetermination cases, focusing specifically on the bombing of Dresden during World War II. For any given target in the bombing, the 30th bomber likely makes no relevant contribution to the destruction—the target has already been utterly destroyed. Even at the point the bomb is dropped it may be clear to the bomber that there is very little chance at all that his bomb will add to the destruction. Yet, the 30th bomber seems just as morally accountable as the previous bombers.

What we need here is a distinction between being held responsible, and being held to account. Many people accept the view, which is highly intuitive, that moral responsibility for x entails causal responsibility for x. That is, one cannot be morally responsible for something if one is not causally responsible for it. If we understand causal responsibility to include negative causation— thus, one can be causally responsible for a harm when one fails to act in such a way as to prevent that harm, in some cases—then this claim, again, seems

---

5. Ibid., 116.

6. See, for example, Frank Jackson, "Group Morality," in *Metaphysics and Morality: Essays in Honor of J. J. C. Smart*, edited by Philip Pettit, Richard Sylvan, and Jean Norman (Blackwell, 1987); Derek Parfit, *Reasons and Persons* (Oxford University Press, 1984). Parfit discusses this in relation to the second mistake in moral mathematics.

highly intuitively plausible. Yet this principle would seem to run afoul of the same sorts of considerations that Kutz raises for IDP. This is because one might think that one can only be held to account for something if one is morally responsible for bringing it about. But this, in turn, buys into a certain assumption—that being blameworthy involves being causally responsible for a harm. And this is what the problematic complicity cases challenge. One can be *blameworthy* with respect to a bad outcome *even if* one did not cause it. The intuition appealed to by many writers on the subject has been to note that if one is a *part of*, a participant in, the cause of the bad outcome, then one is accountable. We don't just evaluate the actions of individuals, we also evaluate collective actions, and we can evaluate the action of an individual as part of a collective. Thus, there are at least two overlapping ways to hold accountable: by judging someone blameworthy, and this can include blameworthy with respect to a harm, and by judging them responsible for a harm. When someone participates in the production of a harm that person's contribution may be large, small, and even non-existent. It is the latter cases that are troublesome when it comes to holding someone responsible for a harm when that person has done nothing to contribute to the production of the harm. However, we can hold the person blameworthy even if not responsible for the harm. We can find some fault with the person even if no harm was produced nor could production of harm be reasonably expected. The best sorts of cases to illustrate this are cases in which someone is a fully superfluous participant and when someone is a bystander to wrongdoing in circumstances in which interference with the wrongdoing would be ineffectual. Truly superfluous agents still seem blameworthy even when their actions made absolutely no causal difference to the outcome—not just a teeny-tiny contribution—even when viewed just as "a part" of the overall cause. Imagine someone who is physically weak and yet tries to help dispose of the body of a murder victim. That person's attempts to help push the body into a river, let's say, add nothing to actually accomplishing the task. However, such a person still seems *complicit*, and blameworthy, even if not responsible. Thus, in providing an account of complicity we need to understand "part" and "participant" in a non-causal way. Kutz's own suggestion on how to understand complicity, in effect, conforms to this model.

His standard of complicity is the following:

> The Complicity Principle: (Basis) I am accountable for what others do when I intentionally participate in the wrong they do or harm they cause. (Object) I am accountable for the harm or wrong we do together, independently of the actual difference I make.[7]

---

7. Kurtz, 122.

Note that the standard advocated by Kutz ties complicity to *intentional* participation in a collective wrongdoing. He locates the difference between his view and the consequentialist view in the following way: his view holds that agents are responsible not for making a causal contribution to a harm, but for the content of their *wills*, regardless of any actual contribution to a harm. Think of a truly superfluous conspirator, who participates in a conspiracy to harm someone and yet ends up making no causal contribution to the harm. Clearly, that conspirator has acted wrongly, is partly accountable for the harm, and is complicit in producing it. One reason for tying complicity to *intentionally* participating in the production of a harm is that it limits the scope of complicity. Without some sort of limit, complicity would be *too* common. It would cheapen the charge of complicity. But, and as Kutz notes himself, this does leave out an important class of cases: bystander complicity cases in which a person witnesses wrongdoing and does nothing to stop it. Such a person may be complicit even though lacking any intention to participate in the wrongdoing.[8]

There is an additional difficulty in tying complicity to a simple intention to participate in a harm all by itself. Imagine the following case:

> Carlene dislikes Dorothy a great deal. Carlene also has very odd views about how the universe works. She believes that if she wants Dorothy to suffer, and she wants it strongly enough, Dorothy will suffer. Others she knows are trying to hurt Dorothy as well—not by sending bad thoughts her way, but by small acts of micro-aggression at work, the cumulative effect of which will be to make Dorothy very sad. Carlene chooses to make her contribution to the effort via the power of what she perceives as her magical thoughts, even though she has no evidence at all that magical thoughts work.

The odd feature of this case is that not only does Carlene's action make no causal contribution, it is the sort of action that *cannot* make a causal contribution. If Carlene is complicit in harming Dorothy, it is not via her attempts at using magic to hurt Dorothy. There are two sorts of complicity that need to be distinguished: *participation* and *tolerance* complicity. This marks a distinction between simply tolerating a wrong—as when one fails to speak up when someone says something bigoted, and participating in a wrong, as when one

---

8. I discuss this issue in more detail in "Kantian Complicity," forthcoming in *Reason, Value, and Respect: Themes from the Philosophy of Thomas E. Hill*, ed. Mark Timmons and Robert Johnson (Oxford University Press).

participates in actively harming someone. What Carlene may be guilty of is a kind of *tolerance* complicity rather than *participation* complicity. She is happy that Dorothy is being harmed, and is not doing anything to stop it. But her efforts to participate are just that—*mere* efforts. There must be a distinction between actually participating and trying to participate. This seems to indicate that for participation complicity, even if one's actions make no causal contribution to the outcome, they are the *sorts* of actions that can make causal contributions to similar outcomes.

In the case of Blake we have someone who is intentionally eating meat and participating in a viciously cruel meat production system—but in eating the meat is she doing something that is itself typically vicious and cruel? So much depends on how the question is framed. In nearby possible worlds it may well be that individual instances of eating meat cause (or raise the expected incidence of) pain and suffering. However, it may be that in nearby possible worlds this isn't at all true for the massive meat production industries, for the same reasons that obtain in our world. Understood that way, she is intentionally participating in the meat production system, though in a way she believes to be causally inefficacious, and she is not performing an action that typically produces bad effects – so that would make her more like Carlene. "Eating meat produced on a massive scale" is like "sending bad thoughts"—it never causes any real harm. Yet there is still a huge difference. One way to account for it is by noting that Carlene is deluded, whereas the person who thinks that eating meat causes animal suffering is mistaken, but not deluded. Blake is complicit—and Carlene not—because it pays to hold Blake accountable. She is in a position to correct the mistake. A person who is genuinely deluded is not in this position. Thus, to be a participant of the relevant sort, the person needs to have capacities that render her amenable to correction.

Why is Complicity Morally Problematic?

Note that Kutz's Complicity Principle is simply a standard. By itself it does not account for the wrongness, it simply helps one pick out instances of complicity.[9] Consequentialists could, in principle, accept everything Kutz presents in the Complicity Principle. Granted, the form of consequentialism would not be straightforward act–consequentialism. A theory that at least incorporated room for policy conformity as a basis for moral evaluation would be able to offer a deeper normative standard. I think that there is some

9. I discuss this further in "Kantian Complicity," op. cit.

truth to this, but, as I hope to show, it is just part of the story. *One* way in which an action can be morally problematic, and thus one way in which we can judge someone to be wrongfully complicit, will involve violations of good producing policies. We can view this as one wrong-making feature. Additionally, what makes the violation one of complicity is that the agent is a participant in a larger project or practice.

Thus, it would seem rule-consequentialist theories would have an advantage here: what is problematic in Blake's case is that her eating of meat is inconsistent with a *policy*, such that if that policy were adopted, it would have enormously good consequences by her own lights. This provides a way to undercut the causal impotency objection. Even if it is true that a single act on a single occasion has no causal impact, as I granted at the outset, a policy surely does. The policy itself introduces another reason for a person to act. Thus, that the policy is a good policy gives me a reason to act according to the policy. Why might this be the case?

Given her normative commitments, Blake must agree that if enough other people stopped eating meat the effects would be very good indeed. Blake is displaying an unwillingness to engage in the cooperative enterprise of ending animal suffering. This is another factor that could render her actions morally problematic. Perhaps, in her case and in the case of any particular individual, the willingness has a purely expressive function: surely she ought to be sad, for example, that the much better system in which animals are not killed for food does not obtain. Eating meat runs counter to this expression—the good, in this case, is the elimination of meat production. In eating the meat she is displaying no negative orientation to meat production. Indeed, she is enjoying it! This explains why, in Blake's case, the natural first reaction is to question her sincerity. The justification she appeals to sounds more like a rationalization than a justification. When one is enjoying the suffering of another that is generally taken to be morally problematic, even if one believes that one's enjoyment does not contribute to the future production of similar harms, and at least an attenuated form of complicity. A person who attended a torture session that would have taken place independently of anything they did, and who enjoyed it, is complicit. The judgment is based on the person's character, not on any contribution they made to *subjective harm* suffered by the person being tortured.

A natural, but mistaken, way to try to unpack this is to hold that persons like Blake are damaging their own character by inuring themselves to the suffering that underlies the industry they take part in. This may be part of the story—indeed, it does seem plausible to hold that the routine eating of meat would undermine any inhibitions one had to eating meat. However, we can imagine that Blake really does care very deeply about animal suffering and would like very much to do things that would stop it. She just, sincerely, does not think that in her case avoiding meat—as long as it is produced in large factory farm settings—really does contribute to animal suffering at all. In her case, her compassionate character is *not* undermined by eating meat. So, what *is* wrong with Blake's *character*? One initially plausible way to go here is to appeal to something that seems empirically plausible about human psychology—that our traits are not really fine-tuned, so that the way that I have described Blake is not psychologically realistic. The animal she eats that she did not cause to suffer resembles other animals. Compassion erosion is likely to be the result. However, this seems too extreme. Imagine another case, in which Blake happens upon the corpse of an animal shortly after it has been struck by a car and killed. It seems very implausible to hold that eating that animal—which certainly resembles animals that have been killed for food— would undermine Blake's character in any way.

Perhaps the best approach is to hold that Blake is not a participant in wrongdoing, but is complicit because she tolerates wrongdoing and even benefits from it. As such she seems to endorse the practice of killing and eating animals, even if she does not in fact endorse the practice of killing and eating animals. There are very good consequentialist considerations in favor of being sensitive to how others might perceive one's actions and the impact those perceptions would have. It contributes to the seeming normalcy of eating meat, and provides cover to those who might feel that something is wrong about it, but still do it since it is ordinary and normal. These are considerations that arise in normal interactions. In many situations this would also count as a wrong-making feature—a consideration that renders the action morally problematic. But, again, we can consider the abnormal case in which everyone is aware that she does not endorse killing animals, and is fully cognizant of her reasoning.

The difference between the factory-farmed meat and the animal killed by the car is that the latter involves accidental, non-intentional, killing. The factory-farmed meat is not produced accidentally, even if it is produced in numbers that render its production practices insensitive to the purchase of individual portions. But we can use a somewhat different case that comes up in the literature: that of the dumpster diver.[10] Colin waits outside a local restaurant every night to take advantage of the meat that restaurant throws away.

---

10. For a discussion of this see Peter Singer and Jim Mason, *The Ethics of What We Eat: Why Our Food Choices Matter* (Rodale 2006), 260 ff.

He realizes that the animals he eats were intentionally killed. And yet, his "dumpster diving" does not seem morally problematic in the same way as Blake's eating the meat she believes will have no causal impact on animal suffering. Perhaps Blake is doing something perfectly analogous, however—we might regard Blake's eating of the meat as just using what is wasted within the meat production system. This seems odd, however, since every single person eating the meat in the factory-farm case can make the same claim to just consuming "waste"—thus, there *is* a disanalogy with the dumpster-diving case. Perhaps it can be captured in the idea that making use of another person's immorality can itself be morally problematic or at least can make one complicit. There may be instrumental worries that one is, in the long term, making the immorality more attractive. One's own use of an immoral practice may not add to that practice, but when others start doing the same, the collective of those who are merely using up what is "wasted" otherwise gets larger, and that collective can have an impact on how much is produced. This allows people to *bypass* norms. A system is established whereby some people habitually benefit from the misdeeds of others, allowing them to reap the benefits without the dirty hands. But this makes the habitual dumpster diver look like he is bypassing norms as well. But if the dumpster diver would be disappointed to see everyone become a vegetarian, and thereby also be unable to benefit anymore from the wrongdoing of others, then he does seem to have a character flaw: a failure to exhibit the right attitude toward wrongdoing.

But I think that, *intuitively*, the best way to capture the difference between cases is to appeal to complicity considerations: Blake is a participant in wrongdoing in a way that Colin is not. When people participate in an activity they *may* be making a joint contribution to an outcome that they all intend to bring about; they may also simply be cooperating with others who are intending to bring about the outcome in question. Blake is participating in wrongdoing by participating in the meat production system. Colin is not. The meat he eats has been thrown away, it has exited the system. There are many different ways in which this complicity can be understood as wrong. Leaving aside the traditional consequentialist worries about the bad effects of this participation, there are also worries about what such participation indicates about the agent's character, some of which can be cashed out in traditional consequentialist terms—such traits that underlie cooperating with evil may systematically produce bad effects, even if they don't in specific instances—but some of which can be cashed out in terms that are consistent with consequentialism though not generally taken to be part of the traditional approach. On a view developed by Tom Hurka, virtue involves having the right kind of orientation or attitude toward good and evil.[11] Such virtue is itself intrinsically good. Though this is not an example that Hurka discusses, one improper orientation toward evil may be that of participation, particularly participation in such a way that seems to endorse a *policy* of participation. On a Hurka type view of virtue this itself would be intrinsically bad, thus, the badness need not be cased out in instrumental terms. Further, as an intrinsic evil, a consequentialist would add these sorts of attitudes to the list of things to be avoided, along with pain and suffering itself. A willingness to participate, even in cases where participation makes no causal difference, is still complicity in evil, and still bad because a willingness to participate in evil displays a bad attitude that is itself intrinsically bad. Thus, when we consider all the value that we might consider, the instrumental effects of character matter, but so does the intrinsic value of the character itself. Eating meat and enjoying it, in Blake's case, displays an inappropriate attitude toward evil—cooperation—even if it in no way, even indirectly, causally contributes to *that* evil. Instead, it constitutes its own form of evil.

However, other appeals to complicity as a way of approaching this problem have been criticized by noting that complicity must involve support of an immoral practice, and one does not support a practice without making a causal contribution to it. On this line of argument, Blake's purchase of meat does not support the meat industry since the purchase makes no difference to the profitability of the industry, and thus Blake does not in fact support the meat industry when she purchases its products. Mark Budolfson writes:

> . . . if the inefficacy objection is correct that when supply chains are long and complex an individual's consumption cannot be expected to make a difference to *the quantity produced*, then an individual's consumption also cannot be expected to make a difference to *the revenues of producers* for the same reasons. This undermines the more general claim of ethical consumerism that by purchasing morally objectionable products one is *complicit in evil* in an objectionable way because one thereby *supports* objectionable firms by *voting with one's dollars* in a way that benefits those firms.[12]

---

11. Thomas Hurka, *Virtue, Vice, and Value* (Oxford University Press, 2003). I don't believe that Hurka's account is exhaustive—there are many virtues that aren't captured by the account. However, I am more convinced now that his account captures an important class of virtues.

12. Mark Bryant Budolfson, "Is It Wrong to Eat Meat from Factory Farms? If So, Why?," this volume.

But this buys into a very narrow notion of complicity. Some may balk at a notion of participation that is non-causal. But even if one holds that participation or support must be causal, one could hold that in cases such as Blake's we have tolerance complicity that involves failure to stand up to evil by a kind of quasi-participation in the practice.

Budolfson also criticizes the complicity approach for being too broad. He uses the example of petroleum products that rely on a petroleum industry guilty of violating moral constraints. Petroleum and petroleum-based products are everywhere. If we are to avoid complicity we would have to give up a vast amount of our "consumption activity," and this is something that most people don't think they are required to do, and don't think their purchases make them complicit in wrongdoing.

> This reveals that almost every consumption activity is complicit in evil in the sense that it depends on and supports companies that violate important constraints to a similar extent that consuming factory-farmed meat does so. But despite all of this, our considered judgment is that it is nonetheless permissible to consume many such everyday products.[13]

This is a standard problem for accounts of complicity. That is, how does one limit the scope of complicity so as to not cheapen the charge of complicity? We distinguish complicity and wrongful complicity. It may very well be that complicity is like causation in that there are causes everywhere—and yet, when we pick out or identify something as a cause, or as the cause, we are guided by pragmatic considerations—such as considerations, in the case of causation, that involve some sort of norm violation.[14] In the case of complicity, there are all sorts of things people are involved in, knowingly, that have some connection to wrongdoing. One thing that impacts complicity is how tight the causal connection is between, let's say, the purchase one is making and the wrongdoing in question, particularly when there is a great deal of intervening agency. Another, and more important for the discussion here, is how many options one has. And yet another, as is the case with causation, has to do with what we consider to be a background condition. In the case of purchasing petroleum products, those products permeate numerous items—it

---

13. Ibid.

14. See my "Attributions of Causation and Moral Responsibility," in *Moral Psychology*, edited by Walter Sinnott-Armstrong (MIT Press, 2008), 423–440.

would be difficult to completely avoid implication in the petroleum industry—that is like a background condition. However, it doesn't mean that someone can't make efforts to minimize support of the petroleum industry. One can make better or worse choices in this regard. Someone who is wrongfully complicit will be someone who has not made these efforts. And there is an important disanalogy with the meat-eating industry, since, when it comes to the meat industry, one can perfectly well avoid supporting it by choosing not to eat meat. In Blake's case, eating the meat is supporting the industry in a situation where there were plenty of other, better, options open to her. What makes her complicit is that she is a participant. What makes that participation morally problematic, in her case, is that the eating of meat displays a willingness to cooperate with the producers of a product that is produced via huge amounts of pain and suffering.

5 IS IT WRONG TO EAT MEAT FROM FACTORY FARMS? IF SO, WHY?

Mark Bryant Budolfson

The signature ethical problem of the global consumer society is our responsibility for the unethical practices that lie behind the products we buy.

—PETER SINGER

What is the best argument against eating meat?[1] One influential argument is that the meat we consume is tainted by factory farming, that this type of farming is the source of the vast majority of the meat that we consume, and that the enormous animal suffering

1. Thanks to Chrisoula Andreou, Derek Baker, Alexander Berger, Heather Berginc, Brian Berkey, Tom Blackson, Cheshire Calhoun, Eamonn Callan, Richard Yetter Chappell, Stew Cohen, Christian Coons, Terence Cuneo, John Devlin, Tyler Doggett, Jeff Downard, Jamie Dreier, David Faraci, Ada Fee, Brian Fiala, Chris Griffin, Liz Harman, Travis Hoffman, Ryan Jenkins, Victor Kumar, Melissa Lane, Alex Levitov, Jonathan Levy, Hallie Liberto, Eden Lin, Zi Lin, Joel MacClellan, Sarah McGrath, Tristram McPherson, Nathan Meyer, Eliot Michaelson, Alastair Norcross, Howard Nye, Ángel Pinillos, David Plunkett, Joe Rachiele, Rob Reich, Ryan Robinson, Julie Rose, Gideon Rosen, George Rudebusch, Carolina Sartorio, Debra Satz, Dave Schmidtz, Dan Shahar, Liam Shields, Sam Shpall, Daniel Silvermint, Peter Singer, Michael Smith, Patrick Taylor Smith, Brent Sockness, John Thrasher, Ian Vandeventer, Chad Van Schoelandt, Alan Wertheimer, Jane Willenbring, Jack Woods, and audiences at the American Philosophical Association, the University of Vermont, Northern Arizona University, the University of Arizona Center for the Philosophy of Freedom, Bowling Green, and the Colorado State University Animal Ethics Conference for helpful discussions. I am especially indebted to conversations with Michaelson, Plunkett, Reich, and Rosen, and to McPherson's arguments in his paper "Why I Am a Vegan," which I follow to varying degrees in a number of places below and which greatly influenced my thinking about these issues. For further illuminating discussion of these issues, in some cases building on the discussion here, see Michaelson's series of posts on veganism and ethics in The Discerning Brute, beginning with "Veganism and Futility," as well as related papers by McPherson, Harman, and Lane. After writing this chapter, it came to my attention that Terence Cuneo also reaches somewhat similar conclusions on the basis of different arguments in his paper "Conditional Moral Vegetarianism."

involved in factory farming cannot be justified by the shallow pleasures that eating meat brings to us, and so eating meat is wrong.

Is this a good argument against eating meat? Here is an initial objection, and an argument for the opposite conclusion: imagine a scenario in which most of the world's corn production is taken over by an evil cartel, which uses slave labor to cultivate and harvest its fields. As a consequence, most of the corn sold in supermarkets is tainted with the suffering and oppression of these slaves. In such a scenario, you would have strong reasons not to consume the corn produced by the evil cartel. However, it doesn't follow that you would have very strong reasons to forgo corn altogether. For imagine that, rather than consume the corn produced by the evil cartel, for an extra dollar per bag you could instead buy corn produced in a humane way by a cooperative of completely ethical local farmers. In such a scenario, even if most of the world's corn would be ethically off limits, there would still be an ethical way for you to consume corn, because you could choose to consume the humanely produced corn instead.

This raises a problem for the initial argument against eating meat above, because similar remarks apply there: even if it is true that most of the meat for sale is from factory farms and thus tainted with suffering that cannot be justified, nonetheless there appears to be another way that you can consume meat ethically: find a cooperative where the animals live a great life, are treated with respect, and are then slaughtered humanely—and buy meat from there. This provides an argument that it's possible for a conscientious consumer to eat meat in an ethical way, at least in areas where humanely produced meat is readily available. Of course, this does not mean that it is easy or cheap: arguably, buying organic is not enough, and unfortunately it may not be financially feasible for low-income families to purchase humanely produced meat. Nonetheless, eating meat in an ethical way is arguably a real possibility for you if you have the luxury of reading this book.

One objection to this argument for the permissibility of eating meat is that even if you buy humanely produced meat you are still contributing to the practice of killing animals in the prime of their lives, which is arguably ethically objectionable. A reply to this objection is that it overlooks the fact that humanely slaughtered animals lead a dramatically better life than they could expect to live in the wild; so, if given a choice, animals would much prefer life on a humane farm to life in the wild, which—according to this reply—means that there is nothing evil or otherwise objectionable about creating and then ending their lives.

Another objection to eating meat is that it is wrong because of the negative environmental impacts of animal agriculture. In reply, it might be alleged that this objection overlooks the fact that if you buy humanely produced meat, a happy side effect is that—according to this reply—you are buying meat that is about as sustainably produced as much of the vegetarian fare that you might otherwise buy from the supermarket instead.[2]

In what follows, I remain agnostic about whether it can be ethical for you to eat humanely produced meat. Many of the other chapters in this volume have a lot to say about that question, and I leave it to them to help answer it. My focus in what follows is on the more specific question of whether it is wrong to consume *factory-farmed* meat from animals that have suffered greatly their entire lives—and more generally, whether it is wrong to consume products that are produced in ethically objectionable ways.

My discussion will be similar to many philosophical discussions of practical issues, which aim to describe the empirical facts that are relevant to an issue, consider particular ethical principles that might be thought to be the correct way of drawing conclusions from such facts, and then consider arguments for specific conclusions based on those facts and principles. Once such an argument is on the table, objections are considered that attempt to show either that the ethical principle it invokes is mistaken, or that it relies on mistaken assumptions about the empirical facts, or that it commits some sort of logical fallacy that prevents the conclusion from following even if the principles and claims are all true. Then, possible replies to these objections are suggested that attempt to show that the objections are misguided, or that a slightly better version of the original argument would get around the objections and establish the same conclusion, and so on. The point of all of this back and forth is to ultimately arrive at the most powerful considerations on all sides of the issue and to make progress in clarifying how to adjudicate the relevant considerations. This is what I started to do in the first several paragraphs above, and it is what I will do in what follows, except in what follows I will occasionally argue for particular conclusions that strike me as true, while trying my best to remain fair and balanced. I leave it up to you to decide whether I'm doing a good job selecting the most important considerations,

---

2. Another reply might draw attention to the *human* welfare effects of eating humane meat versus vegetables and argue that quinoa and many other vegetables including arguably even corn have a surprisingly high harm footprint as a result of the negative effects on human welfare in the poorer nations where it is produced, which result from the rise in prices caused by our collective purchases. (For some representative discussion of the complex underlying issues, see the International Monetary Fund factsheet "Impact of High Food and Fuel Prices on Developing Countries," Joanna Blythman, "Can Vegans Stomach the Unpalatable Truth about Quinoa?," Ari LeVaux, "It's OK to Eat Quinoa.")

and it is of course up to you to decide how ultimately to weigh these considerations and decide what to think at the end of the day.

With that in mind, let's first consider utilitarian arguments that it is wrong to consume animal products from factory farms, such as those offered by Peter Singer. According to Singer, purchasing and eating meat from factory farms is wrong because it has unacceptable consequences on balance for welfare. For example, if I purchase and eat a factory-farmed steak, Singer would claim that my gustatory pleasure is greatly outweighed by the suffering that the cow experiences in order to bring me that pleasure; as a result, Singer would claim that the welfare effects of my eating that steak are unacceptably negative on balance, even if I really enjoy it—and Singer believes that this shows that it is generally impermissible to consume animal products from factory farms.[3]

Although Singer's argument is powerful and initially appealing, there is an important objection to his argument—the inefficacy objection—that claims it is too quick. The inefficacy objection is that even if we agree (as we should) with Singer's premises about the magnitude of animal suffering and the comparative unimportance of gustatory and other human pleasures,[4] his conclusion about the welfare effects of consumption by individuals does not follow,

---

3. Singer, "Utilitarianism and Vegetarianism," and Singer, *Animal Liberation.*

4. It might be objected that, contrary to what Singer assumes, there are in fact good reasons for consuming factory-farmed products on the grounds that we *need* animal products for nutritional purposes, and that factory-farmed products are *cheaper* than the organic alternatives; alternatively, it might be claimed that there are good reasons for consuming animal products that arise from *aesthetic* considerations. However, upon examination, none of these objections to Singer's argument are defensible. For example, consider the idea that there are good reasons for eating meat on aesthetic grounds, because meat is an essential part of "sophisticated" culinary dishes, and so on. The problem with this idea is that it mistakenly assumes that aesthetic experiences that are fleeting, easily replicable, and intellectually insignificant can provide good reasons for torturing animals—which seems false. For example, suppose that a distinctive aroma is released when a particular species of pig is slowly burned alive in an outdoor fire pit, and that some people find this aroma to be "sophisticated" and a good aromatic match for a variety of fine wines. Nonetheless, the prospect of such an insignificant aesthetic experience could not provide good reason to slowly torture a live pig to death in such a way. To put the point another way: while it is arguable (but not obvious) that a pig can be tortured to death if that is the only way to produce a great work of art of everlasting importance, it cannot sensibly be maintained that the shallow and fleeting aesthetics of a fine meal are of sufficient importance to justify such cruelty. Turning then to the idea that good reasons for consuming factory-farmed products arise from nutritional and economic considerations, note first that the evidence strongly suggests that for people like us who are healthy and have easy access to a wide variety of fruits and vegetables, a balanced vegan diet is healthiest and has no important nutritional drawbacks, while in contrast the consumption of animal products is analogous to the consumption of hard liquor, which is unnecessary for our nutritional well-being and actually toxic to our bodies in non-trivial quantities. With that in mind, the fact that factory-farmed

and, upon careful reflection, turns out to be false. That is because an individual's decision to consume animal products cannot really be expected to have any effect on the number of animals that suffer or the extent of that suffering, given the actual nature of the supply chain that stands in between individual consumption decisions and production decisions; at the same time, an individual's decision to consume animal products does have a positive effect on that individual's own welfare.[5] As a result, Singer's premises about animal

animal products are *cheaper* than their organic alternatives does not give rise to weighty reasons for consuming them—for imagine that there are two types of hard liquor: one is produced in a normal, unobjectionable way, while the other is produced using the slave labor of children. If the slave labor variety is cheaper, that does not mean that there is then a weighty reason for consuming the slave labor variety—because there is no ethically important reason for consuming hard liquor in the first place, and so the fact that a particular type of hard liquor is cheaper does not amount to a weighty reason for consuming it, especially when it is produced in a way that is ethically objectionable. For discussion of the nutritional claims made here, I recommend any contemporary public nutrition report prepared by reputable and independent sources (as opposed to a source that is funded or importantly influenced by agribusiness)—for example, the advice of the Harvard School of Public Health is representative:

> The answer to the question 'What should I eat?' is actually pretty simple. But you wouldn't know that from news reports on diet and nutrition studies, whose sole purpose seems to be to confuse people on a daily basis. When it comes down to it, though—when all the evidence is looked at together—the best nutrition advice on what to eat is relatively straightforward: Eat a plant-based diet rich in fruits, vegetables, and whole grains; choose foods with healthy fats, like olive and canola oil, nuts and fatty fish; limit red meat and foods that are high in saturated fat; and avoid foods that contain trans fats. Drink water and other healthy beverages, and limit sugary drinks and salt (http://www.hsph.harvard.edu/nutritionsource/what-should-you-eat/index.html).

Such advice is, quite wisely, designed to be feasible and attractive for its intended audience—but if you read between the lines, it is clear that although such advice allows for the consumption of some animal products in order to maintain its appeal for the intended audience, the science behind the advice suggests that consumption of animal products should be reduced as much as possible, approaching zero. (For provocative further discussion, see for example T. Colin Campbell and Thomas Campbell II, *The China Study* and the references therein.)

5. This is the case at least insofar as individuals are made better off by having their preference to eat meat satisfied. The need for this qualification shows that there is conceptual space for an interesting *paternalistic* welfare-based argument against eating animal products: it could be argued that consuming animal products is wrong because of the negative health-related effects for consumer's *own* welfare. In light of what I argue below, this is a more empirically plausible utilitarian argument than those that rely on considerations of animal welfare, because although an individual's consumption has no significant welfare effects for animals, it clearly has health-related welfare effects for that individual. Unfortunately for Singer, his brand of utilitarianism is not amenable to this sort of argument because he takes the satisfaction of an individual's *preferences* as much more important to welfare than other, more physical and "hedonistic" aspects of well-being. As a result, if Singer's utilitarian theory were modified to make it amenable to this sort of paternalistic argument, then a change in his view would also be required on many related issues in which considerations of paternalism arise, such as euthanasia—for example,

suffering and human pleasures, together with the actual empirical facts about the workings of the marketplace, entail that an individual should expect the effect of his or her decision to consume animal products to be *positive* on balance, in contrast to what Singer assumes. If this inefficacy objection is correct, it undermines the idea that individuals have welfare-based reasons not to consume ethically objectionable products, and shows that Singer's utilitarian principles actually imply that individuals who would do better personally by consuming such products are *required* to do so, which is the opposite of what philosophers like Singer want us to believe.[6]

To make the inefficacy objection a little more vivid, note that everyone can agree that there is a dramatic ethical difference between the following two ways of consuming a T-bone steak: in the first case, a dumpster diver snags a T-bone steak from the garbage and eats it; in the second case, a diner enjoys a T-bone steak at Jimmy's You-Hack-It-Yourself Steakhouse, where customers brutally cut their steaks from the bodies of live cows, which are kept alive throughout the excruciating butchering process. (Once a cow bleeds to death, customers shift their efforts to a new live cow.)[7]

---

the resulting view would then presumably imply that it is permissible to euthanize fully conscious adults whenever their lives are not worth living on hedonistic grounds even when they explicitly insist that they want to continue living, in contrast to what Singer claims (see *Practical Ethics*, pp. 13–14 and 176–178 (3rd edition)). In any event, such an argument is not ultimately plausible: even though there are genuine welfare-based reasons not to eat meat because you would be healthier if you didn't eat meat and your vegetarian lifestyle would influence others in a similarly positive way, those are not strong enough reasons to require you to adopt a vegetarian lifestyle. To see why, note that you would be healthier if you didn't consume alcoholic beverages, and that your abstention from alcohol would also have positive health effects on others, but that does not mean that you are required to give up alcoholic beverages if you really enjoy those beverages and are able to enjoy them without your enjoyment having harmful consequences for others.

6. I examine the inefficacy objection in much more detail in other papers (including "The Inefficacy Objection to Consequentialism" and "The Inefficacy Objection to Deontology"). Among other things, I argue at greater length that an important response due to Peter Singer (and later endorsed by Alastair Norcross and Shelly Kagan) does not succeed. The inefficacy objection has been noted by many authors, although not in connection with the range of related issues discussed here.

7. If this example seems callous at first glance, it may help to note that its purpose is to make salient by analogy some of the horrors of factory farming—for example, some cows are dismembered while fully conscious because of mistakes made in the stunning process at slaughterhouses. Although some such mistakes are inevitable, the actual number of such mistakes is arguably inexcusable, on the grounds that most mistakes could be eliminated by slowing the processing line speed at slaughterhouses to a reasonable level—which would also save countless workers from disabling injuries each year. For a moving discussion of this last issue, see "The Most Dangerous Job" in Eric Schlosser, *Fast Food Nation*.

Everyone can agree that enjoying a steak at Jimmy's You-Hack-It-Yourself is objectionable, whereas enjoying a steak acquired through dumpster diving is far less objectionable, because the welfare effects of eating a steak at Jimmy's are substantially negative on balance, whereas a dumpster diver's consumption has no negative effect on welfare. According to welfarists like Singer, that is the only relevant difference between these two ways of consuming a T-bone steak.

But now consider this question: if you purchase animal products at a supermarket or restaurant, are the welfare effects more like those of buying a steak at Jimmy's or more like those of acquiring a steak through dumpster diving? Conventional wisdom among consequentialist moral philosophers says that the effects are more like eating at Jimmy's; however, the empirical facts suggest that they may be more like dumpster diving, because it is virtually impossible for an individual's consumption of animal products at supermarkets and restaurants to have any effect on the number of animals that suffer and the extent of that suffering, just as it is virtually impossible for an individual's consumption of products acquired through dumpster diving to have any effect on animal welfare. If this claim about the inefficacy of a single individual's consumption decisions is correct, then the upshot is that consuming animal products from factory farms is not objectionable for welfare-based reasons, because there is then no important difference in (expected) welfare effects between an individual consuming factory-farmed products from a store versus a dumpster—and there is nothing wrong with consuming factory-farmed products from a dumpster, as even Peter Singer would agree.[8] (Singer endorses the permissibility of eating meat acquired through dumpster diving on the grounds that such a strategy is "impeccably consequentialist.")

The key empirical claims here relevant to the ethics of consumption are that many products we consume are delivered by a massive and complex supply chain in which there is waste, inefficiency, and other forms of *slack* at each link. Arguably, that slack serves as a *buffer* to absorb any would-be effects from the links before. Furthermore, production decisions are arguably insensitive to the informational signal generated by a single consumer because the sort of slack just described together with other kinds of noise in the extended transmission chain from consumers to producers ensures that significant-enough threshold effects are not likely enough to arise from an individual's

---

8. Peter Singer and Jim Mason, *The Ethics of What We Eat* (US paperback edition) p. 268, and more generally pp. 260–269.

consumption decisions to justify equating the effect of an individual's decision with anything approaching the average effect of such decisions. As a result, for many products in modern society, it may seem empirically implausible that even a lifetime of consumption decisions by a single individual would make any difference to quantity produced and thus the harm that lies behind those products.

For a particularly clear illustration of this, consider the supply chain for American beef. When ranchers who own their own grazing land decide how many cattle to raise, their decisions are sensitive to their own financial situation, the number of cattle their land can support, the expected price of any additional feed that will be needed, bull semen and other "raw materials" that go into cattle production, and the expected price that the cattle will fetch when they are ultimately sold to feedlots. Of these, small changes in the last item—the price that cattle will fetch at the feedlot—are of the least importance, because insofar as ranchers judge that capital should be invested in raising cattle rather than other investments, they will tend to raise as many cattle as they can afford to breed and feed within that budget, letting the ultimate extent of their profits fall where it may at the feedlot. Many ranchers also use the nutritional well-being of their herd as a buffer to absorb adverse changes in market conditions, feeding their cattle less and less to whatever point maximizes the new expectation of profits as adverse conditions develop, or even sending the entire herd to premature slaughter if, say, feed prices rise to levels that are unacceptably high. This serves to shift the ranchers' emphasis in decision-making relevant to herd size even further away from the price of beef. As a result, even if an individual's consumption decisions managed to have a $0.01 effect on the price of cattle at feedlots, the effect on the number of cattle produced would be much smaller than it would have to be in order for the possibility of such a threshold effect to justify equating the expected marginal effect of an individual's consumption of beef with the average effect of such consumption decisions. These facts, together with those that follow, seem to show that there is good empirical reason to think that the actual effect (and expectation) of a single individual's consumption decisions on production is nearly zero and is not to be equated with the average effect of similar consumption decisions across society, contrary to what philosophers such as Peter Singer, Alastair Norcross, and Shelly Kagan have claimed in defending utilitarian arguments against the inefficacy objection.

Furthermore, in the absence of a large shock to the expected price of beef, ranchers who lease grazing land from the government will collectively tend to purchase all of the scarce and independently determined number of grazing permits and raise the maximum number of cattle that are allowed by those permits, because it tends only to make economic sense to hold such permits (rather than sell them to another rancher) if one grazes the maximum number of cattle allowed on the relevant parcels of land. As a result, the number of animals that are raised on land leased from the government appears insensitive to tiny changes in the price of cattle at feedlots.

More importantly, because animal production is so many links in the supply chain away from grocery stores and restaurants, and because each of the intervening links involves waste, inefficiency, and other forms of slack that serve as a buffer to absorb any effect that your personal consumption might otherwise have, it is arguably unrealistic to think that your personal consumption could really have any effect on decisions made at the production end of the supply chain, even when your consumption is considered over the course of an entire lifetime, as noted above. That is because the actual mechanisms by which information is conveyed and decisions are made throughout the supply chain do not seem to give rise to the sort of threshold effects that philosophers tend to imagine as driving the expected marginal effect of an act of consumption toward the average effect of such consumption; instead, waste, inefficiency, and other forms of slack may seem to ensure that the real expected marginal effect of an individual's consumption is essentially zero, because the change in the signal received at the production end of the supply based on a change in a single individual's consumption decisions is almost certainly zero. It does not have a significant chance of giving rise to any tangible expected effects.

Here it may help to focus on the way that decisions are actually made and prices are actually determined at each link in the supply chain, focusing especially on the fact that many of these decisions and price determinations are the result of intuitive human judgment, strategic considerations, and preexisting contracts rather than the result of a frictionless optimization procedure—which means that in practice such determinations are even less sensitive to the noise generated by a single consumer's decisions than they might initially appear. For example, consider the actual human participants at cattle auctions: wearing cowboy boots, standing around in dirt and manure, smoking cigarettes, often distracted and occasionally irrational, and sometimes aiming only to express machismo by means of their bids in the auction. Similarly, consider the actual human participants in production decisions: wearing suits, sitting around in board rooms, drinking coffee, smoking big cigars, often distracted and occasionally irrational, and sometimes doing what is best for themselves rather than promoting the interest of their firms.

As these considerations help make vivid, the actual price and quantity produced are the result of decision processes that have many inputs, and those inputs are arguably insensitive to a change in a single individual's consumption decisions, especially given the actual mechanisms that tend to absorb the signal from a single individual at each stage in the signal transmission chain that lies behind those inputs.

Similar reasoning applies in the case of other animal products, although the relevant market mechanisms are less transparent because of the vertical integration of those industries. Despite that complication, it remains true that the actual mechanisms by which information is conveyed and decisions are made throughout the supply chain are arguably not sensitive to the consumption decisions of individual consumers in the way that would be necessary for there to be important welfare-based reasons for individuals not to consume those products, as with many other products in modern society.

Another important consideration is that even if you would convince many others to be a vegetarian by becoming one yourself, that does not translate into strong welfare-based reasons to become a vegetarian, because even if your vegetarian lifestyle ultimately caused, say, one hundred others to become vegetarians who would not otherwise have done so, their collective consumption decisions might still not have any appreciable effect on the number of animals that are raised and mistreated, because the actual mechanisms in the marketplace may be insensitive to the distributed effects of even one hundred consumers. Of course, this reasoning does not hold true when applied to an influential person like Peter Singer who really does influence enough people to make a difference, but it does hold true when applied to almost everyone else, which means that utilitarianism does not require most individuals to become vegetarians, even if it requires a few influential people like Peter Singer to be vegetarians. For example, just as morality does not require us to act as if we had the talents, influence, and resources that Warren Buffett has, so too morality does not require us to act as if we had the talents, influence, and resources that Peter Singer has.

A related observation is that individual vegetarian acts often have negative unintended consequences that must also be properly accounted for. For example, if I am a vegetarian, I might easily alienate others with my vegetarian acts if they are interpreted as self-righteous, and thus cause others to adopt a policy of never reducing their consumption of meat and never taking vegetarian arguments seriously—and if vegetarians are generally interpreted as self-righteous, that might lead to a consensus among most members of society that vegetarians are radical, self-righteous jerks who should not be taken seriously and who should be scoffed at by others—which then raises the cost of making vegetarian choices for everyone, and is counterproductive in other ways.[9]

So, the inefficacy problem raises an important objection for arguments like Peter Singer's against consuming ethically objectionable products and seems to have a sound basis in the empirical workings of the marketplace. With this introduction to the inefficacy objection in hand, it is useful to consider further potential replies and investigate whether ethical theories that differ from Singer's utilitarianism can offer a more plausible account of whether it is wrong to consume ethically objectionable products.

Perhaps the most common reply is simply to dismiss the inefficacy objection on the grounds that it does not raise any interesting issue beyond the familiar paradox of voting, which asserts that individuals do not have good reasons to vote in elections because there is virtually no chance that a single individual's vote will matter. Unfortunately, this reply seems misguided for several reasons, mainly because insofar as individuals have reasons to vote in elections, there is broad agreement that those reasons arise from one or more of the following considerations:

- the probability that an individual's vote will trigger a *dramatic threshold effect*,
- the fact that voters have a *personal preference* to vote,
- the fact that voters collectively *cause* the outcome of the election in an ethically important way,
- the fact that voters have *non-welfare-based* reasons to vote.

On reflection, analogous considerations seem unavailable to explain why it is wrong to consume products such as factory-farmed meat that are produced in objectionable ways. In particular, an appeal to threshold effects cannot do the job, because, as noted above for empirical reasons inefficiency, slack in the supply chain, and the insensitivity of production decisions to the signal generated by a single consumer seem to ensure that significant-enough threshold effects are not likely enough to arise from an individual's consumption decisions to vindicate an explanation in terms of the possibility of threshold effects. Furthermore, an appeal to personal preferences also cannot do the job,

---

9. Here it may be useful to note that many undergraduate students respond to vegetarian arguments by pledging to eat more meat to "cancel out" the effects of vegetarians.

because most individuals do not have a personal preference not to consume factory-farmed meat and other objectionable products.

Perhaps most surprisingly, it also seems implausible to claim that consumers of factory-farmed animal products *cause* animals to suffer in the ethically important way that voters cause a particular candidate to win an election, because there is not the same kind of causal connection in the animal products case as in the election case. To see why, note that in Australia, New Zealand, and many other large nations consumers have essentially the same animal consumption behavior as in the United States, but such behavior does not cause animals to be mistreated on factory farms rather than treated humanely. The explanation is that the horrible mistreatment of animals on factory farms does not have its proximate cause on the "demand side" in consumer behavior, but instead on the "supply side" in the decisions of producers, as well as in perverse incentives created by irrational government policies. As a result, it seems false to claim that animal consumption *causes* animals to be mistreated rather than treated humanely in a way that is analogous in morally relevant respects to the way that voting causes a particular candidate to win rather than another.

To better illustrate the subtle point here about causal factors, imagine that the United States enacts a general social welfare policy, and when it is implemented some bad consequences result; however, when other nations are examined that have enacted the same sort of policy, we see that such bad consequences do not ensue in those nations, and the explanation is that the US policy was implemented in a corrupt and incompetent way, whereas the policies in other nations were not. This shows that the most relevant cause of the bad unintended consequences in the United States is not the social welfare policy itself, but rather the corrupt and incompetent implementation of that policy—and so it is arguably a mistake to condemn the social welfare policy on the grounds that it was the *ethically relevant cause* of these bad consequences, even if there is a sense in which the policy was a genuine background causal factor and a necessary condition for those bad consequences. By the same reasoning, the consumption behavior of Americans is not the most ethically relevant cause of the bad consequences of factory farming, and does not cause those effects in the same way that voters cause one candidate to be elected rather than another. Instead, the most ethically relevant cause of inhumane treatment of animals stems from the decisions of producers and government, not the decisions of consumers—even though the decisions of consumers are a necessary enabling condition for the bad behavior of producers and government to have the bad effects it does, just as the social welfare policy is a necessary enabling condition for the bad behavior of government implementers to have the bad effects that it does in the example above. Again, this illustrates that it is problematic to claim that animal consumption causes animals to be mistreated rather than treated humanely *in a way that is analogous in ethically relevant respects* to the way that voting causes a particular candidate to win rather than another.

It also seems difficult to appeal to other non-welfare-based-reasons not to consume animals, such as the kind of *complicity in evil* that Tom Regan apparently has in mind in his main argument against eating meat, which is that "Since [animal agribusiness] routinely violates the rights of these animals . . . it is wrong to purchase its products."[10] (Note the close analogy to the argument against eating meat at the very beginning of this chapter.) The problem for this kind of reasoning—even when directed only at factory-farmed meat—is that as just noted an individual's consumption of animals does not seem to *cause harm in the right kind of way*, nor does it *make a difference to the harm* that animals suffer, nor does it even *benefit those who cause such harm*. This last claim may again seem surprising, but the argument for it is that although a single individual's consumption of animal products *does* have a genuine effect on the revenues of *supermarkets and restaurants*, at the same time it does not make a difference to the revenues of *factory farms* for reasons similar to the reasons it does not make a difference to the number of animals produced on such farms: if the inefficacy objection is correct that when supply chains are long and complex an individual's consumption cannot be expected to make a difference to *the quantity produced*, then an individual's consumption also cannot be expected to make a difference to *the revenues of producers* for the same reasons. This undermines the more general claim of ethical consumerism that by purchasing morally objectionable products one is *complicit in evil* in an objectionable way because one thereby *supports* objectionable firms by *voting with one's dollars* in a way that benefits those firms.

---

10. Tom Regan, *The Case For Animal Rights*, updated edition (Berkeley: University of California Press, 2004), p. 351. Tristram McPherson's view in "Why I Am a Vegan" (unpublished manuscript) is similar to Regan's but more clearly and fully developed; my objections here also apply to McPherson's view. McPherson is developing a response to my arguments in important current work, and at this point it is unclear to me to what extent we will ultimately disagree and to what extent our views will ultimately converge. As I note elsewhere, I am indebted to McPherson's discussion, which I follow to varying degrees in a number of places here, and which have greatly influenced my thinking about these issues, as well as providing the initial impetus for all of my thinking about these issues.

As additional confirmation that there is something wrong with quick invocations of complicity in evil such as Regan's, note that these attempted explanations overgeneralize and imply that you are almost never permitted to consume anything at all, because petroleum companies routinely violate significant constraints,[11] and almost every possible consumption activity depends on and supports such companies to a much greater extent than the activity of buying animals at supermarkets and restaurants depends on and supports factory farms.[12] For example, if such a simple notion of complicity in evil really did give rise to strong reasons not to consume products, then it would be wrong to consume petroleum products because of the oil industry's complicity in serious harm, and it would be wrong to consume almost everything else as well, because almost everything depends on petroleum products via dependence on transportation companies, which turn a blind eye to oil companies' abuses that are known to benefit transportation companies in the form of lower fuel costs. This reveals that almost every consumption activity is complicit in evil in the sense that it depends on and supports companies that violate important constraints to a similar extent that consuming factory-farmed meat does. But despite all of this, our considered judgment is that it is nonetheless permissible to consume many such everyday products.[13]

How then can we explain the ethically relevant differences between consuming products that are produced in an ethically objectionable way and those that are not? After all, most people would agree that even if what you consume as a single individual really makes no difference, there are still some *particularly objectionable ways* of being connected to evil that are impermissible. This, then, leads to the philosophical question of how exactly to distinguish the particularly objectionable ways of being connected to evil from the relatively innocuous ways of being connected to evil. If we can identify a compelling account of this distinction and see how it applies to a variety of cases, such as the consumption of factory-farmed animal products, we can then

---

11. For examples of routine abuses, see Peter Maass, *Crude World*, especially chapters 2, 3, and 4.

12. Note especially that individual purchases of gasoline for personal use are often permissible even though we thereby purchase gasoline directly from the petroleum companies themselves, or are at least only one step in the supply chain removed from such companies—and in the gasoline supply chain there is much less of a buffer caused by waste and inefficiency than in the supply chain for animal products.

13. I discuss other possible responses to the inefficacy objection in much greater detail in other papers, as referenced several notes above.

determine whether that account delivers a compelling and satisfying package of theories and verdicts on the cases we care about. If it does, then we will have answered the main philosophical questions that arise from the issues discussed in this chapter.

One intuitively appealing way of making the distinction between permissible and impermissible connection to evil is to invoke the notion of *the degree of essentiality of harm to an act*, claiming that, for example, consumption of a product is particularly objectionable the more essential it is to that product that harm or the violation of rights lies behind it. To illustrate this basic idea, consider a can of vegetables sold at a supermarket that is produced in a normal way. Although the production of those vegetables might depend on petroleum products and thus involve a surprisingly high footprint[14] of harm and connection to evil, it is nonetheless *highly inessential* to that product that such evils occur in the background, because there is nothing in the nature, actual production, actual consumption, and so on of that product that necessitates harm or the violation of significant rights. This provides a principled reason for explaining why you do not have strong reasons not to consume a can of vegetables even though you know that they have a surprisingly high footprint of harm because of the petroleum products that ultimately lie further behind their production and distribution and are produced by corporations that often violate significant constraints. As a result, invoking this idea allows for a principled distinction between connections to evil that seem innocuous and connections that are not, and avoids overgeneralizing and implying that it is impermissible to consume everything, even a can of corn (as the overly quick appeal to complicity in evil overgeneralizes).

Setting aside for a moment the precise details of this notion of *degree of essentiality*, for current purposes it is worth noting that it is embedded in moral common sense, and as a result it seems essential to an intuitive explanation of the relevant cases, especially in light of the failure of more familiar

---

14. The *footprint* of an act of a particular type is simply the *average effect* of all actual acts of that type for some particular kind of effect. I discuss the relevance of footprints to ethics and public policy in my paper "Collective Action, Climate Change, and the Ethical Significance of Futility," where I argue that they are overemphasized and sometimes mistaken guides to what should be done, and that ethical reasoning that depends on footprints generally commits what I call *The Average Effects Fallacy*, which is the fallacy of equating the ethically relevant effects of a particular act with the average effects of all of the actual acts of that type. I suspect that consequentialists who quickly dismiss the inefficacy objection are often tacitly committing a version of this fallacy.

ethical notions to adequately explain those cases. For an example of its use in moral common sense, consider the ethical view that is probably endorsed by most actual vegetarians, which is suggested by one interpretation of this quote from Michael Pollan:

> Like any self-respecting vegetarian (and we are nothing if not self-respecting) I will now burden you with my obligatory compromises and ethical distinctions. I'm not a vegan (I will eat eggs and dairy), because eggs and milk can be coaxed from animals without hurting or killing them.[15]

There are two ways of understanding the underlying ethical principle here. On one interpretation, the idea is that consuming animal products is permissible only if those products do not *actually* have any footprint of harm or killing. However, that cannot be the interpretation of this principle that is endorsed by most actual vegetarians, because most actual vegetarians are ovo-lacto vegetarians who believe it is permissible to consume factory-farmed dairy products—even though those products have a very high footprint of harm.[16] This suggests that most actual vegetarians interpret this principle in a way that leans heavily on the idea that even factory-farmed dairy products *can* be coaxed from animals without hurting or killing them. Presumably, the idea here is that even if factory-farmed eggs or cheese have a disturbingly high footprint of harm, nonetheless it is highly *inessential* to those products that such harm lies behind them, and as a result consuming them does not connect one to harm in a way that is impermissible. So on this second interpretation, the crucial ethical issue is whether a particular animal product *can* be produced without harm—and taken at face value, this is subtly different than the issue of whether a particular product *actually is* produced without harm. And as we've just seen, this second interpretation and its invocation of the degree of essentiality of harm seem deeply rooted in the moral thinking of most actual vegetarians.

As further confirmation of this, note that alternative interpretations of most vegetarians' underlying ethical principle would imply, contrary to

---

15. Michael Pollan, *The Omnivore's Dilemma*, p. 313.

16. Note that such a principle is implausible because (as explained above) it overgeneralizes and implies that it is impermissible to consume almost everything in contemporary society, since almost every product has a surprisingly high footprint of harm or killing.

their view, that eating eggs and dairy is generally *more* objectionable than eating beef, because the actual suffering of laying hens and dairy cows on factory farms is far more extreme than the actual suffering of cattle raised for slaughter, given that most cattle raised for slaughter tend to be raised in good conditions on ranches, and only encounter factory-farming operations when transported to feedlots and then to slaughterhouses—and even then, significant suffering tends to be visited probabilistically on only some of the cows. This is in contrast to the laying hens and dairy cows that are used to produce factory-farmed eggs and dairy products, which experience the worst treatment of any animals in contemporary agribusiness. As a result, the harm footprint associated with each calorie of energy from factory-farmed eggs or dairy is generally higher than the harm footprint associated with beef, and similar remarks apply to other measures of ethical objectionability that would be relevant to other familiar ethical theories; nonetheless, most ovo-lactovegetarians believe that eating meat is wrong, but that consuming factory-farmed eggs and dairy is permissible, which is why familiar ethical theories do not provide a charitable interpretation of their view. Instead, the most charitable interpretation of their view is that what matters is whether it is *highly essential* to an animal product that harm or killing lies behind it. The idea of ovo-lactovegetarians is, presumably, that it is not essential to eggs and dairy that harm or killing lies behind them, whereas it is quite essential—given actual facts about cause and effect and facts about technological possibility and feasibility in the actual world—that killing lies behind eating meat; therefore, eating meat is impermissible, whereas consuming eggs and dairy is permissible, even if the harm footprint of the latter is greater than the harm footprint of the former.

Setting aside the views of actual vegetarians, further confirmation of the explanatory power of this notion of degree of essentiality is provided by consideration of other cases. For example, suppose you learn that a computer you are interested in purchasing is made from metals and other inputs that are themselves produced in a way that is objectionable. Although that harm might be serious, it might also be true that there is no practical way that the computer manufacturer can do anything about it, and it is in no way central to the design or functioning of those computers that such harm lies behind them. In such a case, buying the computer need not be impermissible even if some action that lies far behind them is impermissible. This is in contrast to a different case involving the same harm footprint in which it is *fairly essential* to the computer that such harm lies behind it, perhaps because of some engineering decision that requires it to be produced in a way or from materials that involve such harm.[17]

Why would morality make such a distinction between ways of being connected to evil? If there were no good answer to this question, then we should doubt whether a principle that invoked the degree of essentiality of harm was a genuine moral principle, even if it seemed to correctly capture our initial intuitive judgments about cases. However, on further reflection it is not that surprising that morality would make such a distinction, as it seems to be a consequence of the more general compelling idea that morality distinguishes between what is within a single individual's control and what is not. If it is highly inessential to an action open to you that harm lies behind it, but at the same time background actions by others that you cannot change and that are far removed from any direct connection to any action of yours would give that option a high harm footprint, the current proposal is that morality does not assign that as much negative weight as if a similarly high degree of harm would be directly caused by your choosing an option, or if it is relatively essential to an option that a similarly high degree of harm lies behind it. This is merely one way that morality distinguishes between relevant facts that are within a single individual's control and facts that are not.

In sum, the discussion above suggests that the best explanation of what consumers are required to do when products are produced in morally objectionable ways is more subtle than it initially appears. Among other things, it suggests that typical appeals are too quick to the welfare effects of consumption or connection to evil practices that lie behind our products, and do not invoke principles that are ultimately defensible. At the same time, appealing to the *degree of essentiality of harm* seems to allow a better explanation of the cases that we care about and remains consistent with the compelling idea that, for example, the most decisive fact relevant to the ethics of eating meat is that a person's gustatory pleasure is of little ethical significance compared to the

---

17. As another example, consider that the law makes a similar appeal to the degree of essentiality of harm. For example, in *New York v. Ferber* and in *Ashcroft v. Free Speech Coalition* the US Supreme Court held that a compelling state interest exists to prohibit the promotion and consumption of child pornography insofar as that pornography is "intrinsically related to the sexual abuse of children." This appeal to the notion of the degree of essentiality of harm that lies behind sexually explicit materials involving children provides an important part of the court's basis for distinguishing between, on the one hand, objectionable child pornography that may be constitutionally prohibited by legislation (e.g., actual videos of child sex acts, where it is *highly essential* to the pornography that harm lies behind it), and on the other hand, constitutionally protected and arguably unobjectionable depictions of children engaging in sexual acts (for example, drawings in textbooks, or depictions by actors in fictional films).

suffering that animals must experience in the service of that pleasure. And arguably, this view implies that there is something that is genuinely more objectionable about eating meat from factory farms than eating humanely raised meat. The general philosophical point is that facts about the pleasure that we get from products and facts about the harm that lies behind those products do not lead to conclusions in the simple way that utilitarian reasoning might initially suggest, but must instead be marshaled into more subtle arguments. As these subtleties are clarified, many of our prior judgments will be vindicated—but a few may also have to be revised.[18]

---

18. For further discussion of the ethics of collective action and some prior judgments that may need to be revised, see my paper "Collective Action, Climate Change, and the Ethical Significance of Futility." The current chapter is intended for a general audience. I discuss other possible responses to the inefficacy objection and related issues in much greater detail in other papers, as referenced several notes above.

6 POTENCY AND PERMISSIBILITY

Clayton Littlejohn

Introduction

Smoking is pleasurable. It isn't just pleasurable; it is *immensely* pleasurable. A smoker can derive more pleasure from a single cigarette than, say, most of us could derive from eating a steak. Not only that, smoking is *distinctly* pleasurable. There is no good substitute for smoking a cigarette. When you want the pleasures associated with smoking a cigarette, only a cigarette will do. While pleasure is unarguably a contributor to your well-being (and arguably the only thing that contributes directly to your well-being), nobody would argue that you ought to smoke in order to maximize well-being. Isn't this paradoxical?

No, it isn't. It isn't even mildly puzzling. Even if hedonism is true and your concern is only with your own well-being, you'd have to take account of the pleasures you'd derive from smoking and then weigh them against the pleasures you could attain if you quit or decided against starting. Even if you don't get lung cancer from smoking, it can shorten your life or interfere with your ability to enjoy an active lifestyle. There's a very good chance it will harm you by causing you to undergo something rather unpleasant, but there's also a very good chance that it will harm you by robbing you of pleasures you could have experienced otherwise.

This isn't a chapter about smoking. Philosophers don't write about smoking unless they're taking money from a tobacco company. Sadly, I'm not. This is a chapter about the unreflective carnivore. Most of the people near and dear to me (myself included) are unreflective carnivores. We act as if we don't see any good reason not to order meat in restaurants or buy it in grocery stores. If you press us, we'll say that we eat meat because it brings us immense and distinctive pleasures. We seem to be under the impression that we maximize our own well-being by eating the significant quantities of meat that we do.

This kind of thinking is as confused as it is common. People who see that an argument from hedonism in favor of smoking is muddled should know better than to think that acting like the unreflective carnivore is a good strategy for maximizing well-being. The harms to health caused by consuming the amount of meat consumed by the typical meat consumers in the United States or Great Britain aren't negligible.[1] Moreover, if we're honest with ourselves and we think about the amount of pleasure we derive from eating steak, we cannot really say that eating steak is *immensely* pleasurable. The pleasures associated with eating steak aren't like those associated with smoking much less like those associated with taking cocaine or Ecstasy. My guess is that once you cultivate a taste for vegetarian cooking, you wouldn't rate your vegetarian meals lower than, say, your steak or ribs. The only reason people think otherwise is that they haven't really bothered to test this out for themselves.[2]

This isn't a chapter about whether cultivating a taste for vegetarian cooking is a good strategy for maximizing well-being. Though such cooking is likely a good strategy for doing that, I probably cannot convince you that it is unless you test my hypothesis by developing a taste for vegetarian cooking and doing some research on the health benefits of a vegetarian lifestyle. This is, however, a chapter about the unreflective carnivore, someone who makes decisions about what to purchase on the basis of a yen, a craving, or mistaken belief about what maximizes their own well-being. The question is whether there's any justification for acting like the unreflective carnivore.

---

1. Of course, it would be a stretch to say that prudence requires a strict practice in which you'd refrain from eating meat. Your ability to do what's prudent might depend upon whether you try to stick to a strict rule, but that's another matter. The case against eating meat isn't a prudential one.

2. An anonymous referee thinks that I'm overconfident in my judgments about taste. My judgment is based on my experience and the experiences of others. This is, however, an empirical matter and we should either be guided by our own observations or the observations of others. Readers who are skeptical of my claim are free to conduct their own experiments.

If the unreflective carnivores near and dear to us are asked to reflect, they'll admit that they wouldn't set dogs on a rabbit and wouldn't set fire to a chicken. They wouldn't run a lamb down with their car. They wouldn't shoot a dolphin from the deck of a boat. When they watch nature shows, they are horrified if they see a baby elephant killed by lions, and they get misty when they watch the desperate mother elephant search for her young. In spite of this, they'll order rabbit, chicken, or lamb in a restaurant. They think that it matters, in some sense, what happens to animals, but they think that there's a sense in which their actions don't have any moral significance. I want to see if these *no-difference* defenses hold up to scrutiny.

It's important to fix some parameters for this discussion. The people near and dear to me live in the United States and Great Britain. They can afford to eat in restaurants and purchase meat in the grocery store. They can also afford to eat fresh fruits and vegetables, which happen to be in abundant supply where they live. If anything, they'd save money by becoming vegetarians. Although such a diet isn't appealing to them, they know enough about cooking that they could easily learn to cook delicious and healthy vegetarian meals if pushed to do so. They don't suffer from any bizarre medical conditions that would require a daily dose of bacon. They aren't utility monsters. They don't derive freakish amounts of pleasure from eating steak or chicken pot pie. When I've served them vegetarian meals, they haven't rated them lower than they'd rate a chicken or pork dish that I'd prepare or that they'd prepare. They know that the animals they eat are sentient, capable of feeling pleasure and pain. They believe that some of the animals they eat lived reasonably good lives and were killed humanely. They believe that some of the animals they eat lack complex mental lives and so cannot have self-regarding desires about their own futures. They also realize that the animals they eat are typically killed in vast numbers so that the decision to kill a chicken or lamb, say, isn't directly in response to the order they place in a restaurant or a purchase they make in a market. This should give you some sense of the kind of person I have in mind when I evaluate the no-difference defenses.

Let me fix one more parameter. The people near and dear to me aren't neo-Cartesians and so they agree that some of the animals they eat are capable of suffering as well as capable of enjoying pleasurable experiences. They agree that this means that some of the non-human animals we eat are moral *patients*, creatures who have interests that have non-instrumental moral significance. They might not describe their view in these terms, but this is a view that they're committed to and something they'll accept without too much of a fuss.

To see why they should accept this, let's review the marginal cases argument.[3] The starting point of the argument is this:

> MC1: All living, sentient humans are moral patients.

That is to say, all living, sentient humans have interests that carry some sort of moral weight. It can be *prima facie* wrong to do something that harms those interests or fails to promote those interests. The next step is this:

> MC2: If all living, sentient humans are moral patients, each of these individuals must possess some set of features that's sufficient for conferring the status of moral patient upon them.

The idea here is that for any being that has the status of being a moral patient, there is some feature or non-empty set of features that's sufficient for conferring that status upon them. Those who accept (MC2) insist that there's some reason why moral patients have that status and that the possession of this status will be connected to the features of the individuals that possess it. If you accept this, then it's hard to see how you could avoid saying the following:

> MC3: Any set of features that's sufficient for conferring the status of moral patient upon the living, sentient human animals that have them would be sufficient for conferring the status of moral patient on any non-human animal that has this set of features.

Moreover, it seems that the following is quite plausible:

> MC4: There are non-human animals that we eat that have these features.[4]

If this is correct, we have our conclusion:

> MCC: There are non-human animals that we eat that are moral patients.

---

3. See Singer (1993) for an early discussion of the argument.

4. An anonymous referee noted that this premise, as stated, is quite abstract. I see this as a virtue, not a vice. Readers are free to plug in the features that they think matter to personhood and see whether it's true that these features allow us to say that the class of moral patients includes all living persons and none of the animals we enjoy eating. I simply wanted readers to know the shape of the challenge that the marginal cases argument presents. I think the strongest objection to the use of that argument in arguing for ethical vegetarianism is in the transition from the argument's conclusion to further claims about what we should do in light of (MCC).

Once this conclusion has been established, we have to acknowledge that there are non-human animals that have a moral status that ground obligations. If you're a consequentialist, you'd have to acknowledge that the considerations about aggregate well-being that determine the rightness of your actions will include considerations about the well-being of some non-human animals. If you're a non-consequentialist, you'd have to acknowledge that there are principle-based protections that apply to non-human animals that are similar to those that we recognize for humans at the margins of life. It will matter later that the argument doesn't simply establish that animals matter morally, but that they matter morally in particular ways that humans do.

If readers don't want to accept the argument's conclusion, so be it. Objections to this argument have been dealt with elsewhere and my unreflective carnivores don't try to undermine this argument to defend their behavior.[5] Still, a brief word about (MC4) is in order. In offering this argument, I didn't say that there aren't significant moral differences between puppies and children. There might well be significant moral differences between all living, sentient humans and all non-human animals. That doesn't matter. All that matters is whether these differences are sufficient to undermine the argument that fish, chickens, cows, ducks, rabbits, pigs, lambs, buffalo, and so on have *some* kind of moral status. Think about human infants and three ways that a human infant might be:

a. Sentient, capable of having self-regarding desires about its future, capable of developing into an adult with normal intellectual and emotional capacities.
b. Sentient, capable of having some self-regarding desires about its future, but incapable of developing into an adult with normal intellectual and emotional capacities.
c. Sentient, incapable of having some self-regarding desires about its future, and incapable of developing into an adult with normal intellectual and emotional capacities.

If you thought that only infants with the characteristics described in (a) had moral status, you could try to resist the argument for (MCC) on the grounds that you thought that the non-human animals we eat don't have the features necessary for moral status. The trouble with your position is that you could

---

5. See Kazez (2010) and Norcross (2004) for discussion of objections to the marginal cases argument.

only deny (MC4) by excluding many infants from the moral sphere. That's monstrous. On the view you'd be defending, there would be healthy infants that weren't moral patients. Since they wouldn't have any interests that carried any moral weight, your view would be that there wouldn't be any principled objection to eating them or using them for parts. (At least, there wouldn't be any such objection that appealed to the status possessed by such infants since your view would be that they didn't have any.) If, on the other hand, you stand rightly opposed to eating all human infants by virtue of the fact that all of them have moral status, you couldn't resist (MCC) unless you thought that the non-human animals we ate weren't sentient. You don't think that. Insofar as you think that sentience is sufficient to confer moral status upon all sentient infants, you should think that it is sufficient to confer a kind of moral status upon the sentient animals we're eating.

The First No-Difference Argument

If the marginal cases argument is sound, you shouldn't be terribly impressed with our first no-difference argument:

> I'd agree that you shouldn't eat things like monkeys, dolphins, or dogs because they have the higher capacities needed to form self-regarding desires about their futures. When we're dealing with creatures like that, we're dealing with creatures that can be harmed by being killed because their futures can be good for them. When we're dealing with simpler sentient creatures incapable of thinking about themselves in the future, however, their deaths cannot be bad for them because they cannot have any self-regarding desires about what happens to them in the future. So long as they're humanely housed and killed, no wrong has been committed. Thus, it makes no moral difference whether we eat sentient creatures or not, provided that they lack high-order cognitive capacities involved in forming self-regarding desires and they were killed humanely.

The argument isn't the slightest bit convincing when it's applied to infants that lack the potential for developing the capacities necessary for having self-regarding desires about their own futures, so it shouldn't be any more convincing when it's applied to creatures like fish or chickens.[6] On the most

---

6. Assuming, of course, that chickens and the relevant fish are sentient. For further discussion of deprivation and desire-based accounts of well-being, see Bradley (2009).

plausible account of why our deaths are bad for us, deaths are bad for us by virtue of depriving us of something valuable; death is bad by virtue of the fact that it deprives us of a future period of existence that would have contained a positive level of well-being. This account explains why the death of disabled infants can be bad for the infant and it applies to sentient non-human animals, too.[7]

Before pressing on, there is a point that I should clarify. I don't think that *this* is how we should respond to the first no-difference argument:

Death's badness is due to the fact that it deprives something of a future in which it is a sentient creature that lives a life that is overall good for it. Thus, it is wrong to kill sentient animals if they could have continued living lives that were overall good for them.

This response moves too quickly from a claim about the kinds of events that can harm an individual to a claim about what sorts of events it would be wrong to cause. This response assumes, implicitly, that causing a deprivation involves *harming* and *wronging*, and I think we should leave it open whether harms are invariably wrong.[8] The deprivation account of death's *badness* seems to imply that an earlier death is typically going to be a worse death, at least if we're assuming that the relevant individual stands to live a life that's overall and uniformly good. If the *badness* of a death (which is measured in terms of the magnitude of a deprivation) is a sufficient ground for the *wrongness* of causing death, the deprivation account of death's badness implies that it is wrong to terminate a pregnancy when bringing the fetus to term would have resulted in a life that was overall good for the individual. Indeed, it seems to imply that there's a wrong in the case of abortion and infanticide that's roughly equivalent.

The way to avoid this implication is to draw a distinction between the deprivations that cause harm or leave someone worse off and the deprivations that do so in such a way that's wrongful. If we assume that all deprivations are harms, we shouldn't assume that it is invariably wrong to cause them. (If we

---

7. It's an interesting question whether non-culpable ignorance of the fact that sentience is sufficient for moral status can subvert obligation or exculpate. For arguments that it cannot, see Arpaly (2004), Harman (2011), and Littlejohn (2014). For arguments that it can, see Zimmerman (2008).

8. This kind of distinction is crucial for understanding Thomson's (1971) defense of abortion and nicely explains why the austere basis that Marquis (1989) tries to argue from doesn't support his conclusions about abortion.

assume that it's always wrong to harm, we shouldn't assume that all deprivations harm. I'd prefer to speak as if deprivation and harm are the pair that's
more intimately connected.) Here's one reason to draw this distinction. If we
think that an infant once was a fetus (i.e., that the patient that is the infant
was once identical to a fetus), we'd have to say that terminating a pregnancy
causes a deprivation to the relevant patient that's equivalent in terms of the
magnitude of harm to the deprivation caused by an act of infanticide. And if
we assume that the badness of the deprivation is itself the sufficient ground of
the wrong of infanticide, it seems we get the result that aborting a pregnancy
is no less wrong than infanticide because it's no less worse for the relevant
individual.

It might initially seem odd that there's a gap between causing a depriva-
tion/harm on the one hand and wrongdoing on the other, but examples can
help to make this intuitive. Suppose we were to discover that planaria had
some minimal degree of sentience. You'll recall that if you divide a planarian
into two parts, each part will regenerate the parts needed so the result will be
two living flatworms, not one dead flatworm in two pieces. Phyllis the planarian is about to be cut into two pieces but you intervene so she suffers only a
slight knick, not a complete separation. She heals so we don't have a case with
one dead planarian or two living planaria. I don't think that the parts of Phyllis that could have grown into full planarian had an interest in being split off,
but there would be two individual planarian that would have been better off
as a result.[9]

Of course, someone might ask at this point why we should think that the
animals that cannot anticipate their own futures are patients in the sense that
they have interests that carry moral weight. The answer is easy. Although
there's a potential gap between harming and wronging, we don't think that
the possibility of such a gap allows us to terminate the life of an infant in
groups (b) or (c) discussed on page 103. The marginal cases argument gives
us some reason to think, then, that the mental capacities common to these
infants and the relevant animals are sufficient for conferring upon both the

---

9. For the purposes of the example, I'm assuming that if there's an individual that has some
positive level of well-being (e.g., a sentient planarian that results from splitting Phyllis into two
pieces) we can say that this individual would have been harmed if we had done something that
resulted in a situation in which it isn't true that that individual exists and enjoys at least as
much well-being. I appreciate that this is controversial, but I don't think anything I defend here
turns on this. The example is a helpful reminder of the gap between deprivations and wrongs.
See Kamm (2005) for further discussion of this point in connection with recent debates about
the use of embryos in medical research.

kind of moral status that makes it seriously wrong for us to harm them by killing them.

To understand the moral distinctions between non-creation, abortion, and infanticide, we shouldn't just look to considerations having to do with the aggregate well-being of non-creation, abortion, infanticide, and the relevant feasible alternatives. We can distinguish abortion from infanticide (and contraception) by pointing out that it's only in one of these cases that we're dealing with a moral patient. It isn't in the interests of the fetus to develop into an infant, you could say, just as it isn't in the interests of an egg or sperm to fuse with the other. The account that I'd recommend combines a deprivation account of harming with a patient-affecting restriction on wrongdoing. It should be noted that the account is incompatible with any consequentialist approach to right action that accepts totalism:

Totalism: The morally relevant outcome of an action is the possible world that would be actual if the action were performed.[10]

If you accept totalism, you wouldn't think that there would be much difference between deciding against bringing a patient into being and snuffing the patient out. This, however, is an important distinction if you accept the patient-affecting restriction, as the restriction says in effect that only the snuffing is a violation of the principle of non-maleficence. There might be consequentialists that reject totalism, but it's hard to see how they could defend such a move. If you think that the good is prior to the right, it's hard to see how you could sensibly distinguish the relevant goods from the irrelevant goods by appeal to considerations like whether the relevant goods would be enjoyed by actual agents as opposed to merely possible agents you could bring into being. If we think of morality as primarily concerned with how patients are treated and only derivatively concerned with things like aggregate levels of well-being, the restriction might seem intuitive and consequentialism should appear to be problematic.

The Second No-Difference Argument

The first no-difference argument fails because it tries to exclude certain sentient creatures from the moral sphere. The marginal cases argument should

---

10. See Carlson (1995: 10).

remind us of two things. The first is that when you raise the bar, you run the risk of excluding some living humans from the moral sphere. The second is that if we draw fine-grained distinctions between kinds of moral status, we'll find that some sentient non-human animals will have the same kind of moral status as some living humans. We shouldn't just think that animals are potential victims. We should also recognize that this potential for victimhood grounds obligations concerning animals that are similar to the obligations we have concerning fellow humans.

The second no-difference argument is supposed to avoid these worries because it doesn't aim to show that there are sentient animals that lack moral status. Instead, it's supposed to show that the unreflective carnivores often don't have the obligation to refrain from buying or ordering meat because of the causal structure of the supply chain:

> In sitting down at a restaurant that has lobster on the menu, it doesn't matter what I order because if I don't order the last lobster someone else will.[11] Thus, while there might be obligations not to harm animals, the existence of such obligations has nothing to do with me on this occasion.

The argument has limited import, of course, because it only applies to those cases where a living animal would be killed if someone were to request it. We'll discuss cases where the animals have already been slaughtered and prepared for consumption in the next section, so let's just focus on cases with this structure for now.

If the argument shows that the unreflective carnivore has no responsibility to refrain from eating the lobster, it also seems to show that the unreflective carnivore cannot be held responsible for deciding to order the lobster. Although it's possible to fail to meet your responsibilities without being responsible for the failure, there are nevertheless principled connections between responsibility in its forward- and backward-looking sense. If you *know* that you're under no responsibility to X, you cannot be held responsible for X-ing. Thus, if the argument shows that the unreflective carnivore knows that she's under no obligation to refrain, it also shows that the agent isn't culpable for her decision.

These results strike me as problematic. Even if the unreflective carnivore knows that another customer would order the lobster if she didn't, she would still be under an obligation not to order it. (Assuming, of course, that she'd be under this obligation if she were the restaurant's only patron.) If you like

---

11. Gardner (2007: 72) dubs this kind of argument the "arms dealer" argument.

slogans, here's one: preemption doesn't bear on permissibility. Moreover, even if the unreflective carnivore knows that the lobster is going to get it one way or the other, she's still culpable if *she* is the patron who orders it. The question is how can we make sense of this in light of the fact that the lobster won't be made worse off than she would be otherwise by the agent's decision to order her for dinner.

Let's start with something simple. Suppose you think that there's a principle of non-maleficence, one that says that it would be *pro tanto* wrong for you to harm a moral patient.[12] If you order the lobster and preempt another patron waiting in the wings, you cause the lobster's death (with help from the restaurant staff, of course). So, *you* harm the lobster. So, you commit the wrong, right?

One problem with this line of thought is that it seems susceptible to the following line of objection:

Look, you couldn't have harmed the lobster if the lobster isn't harmed as a result of your action. For the lobster to be harmed as a result of your action, it has to be that the lobster suffers a harm. This requires that the lobster is worse off than she would have been otherwise. But the lobster isn't worse off than she would have been otherwise as a result of your action. If you didn't order the lobster, someone else would have. Since there was no harm, there wasn't any violation of the principle of non-maleficence.

It seems rather plausible that an agent doesn't harm a patient in our example if the patient doesn't suffer any harm by being killed. I certainly don't want to defend the view that you can harm the lobster even when the lobster isn't harmed. What should we say in response?

The objection assumes that harm should be understood *comparatively* in accordance with this counterfactual account of harm: an event is harmful for an individual if things go worse for this individual than they would have gone if the event had not occurred (Bradley 2012: 396).[13] Applied to the case at hand, you don't harm the lobster by ordering it and thereby ordering its death

---

12. See Ross (1930). Dancy (1998) provides a helpful discussion of the differences between Ross' understanding of harm and the duty of non-maleficence and the consequentialist understanding of the duty.

13. There are non-comparative accounts of harm to consider. See Harman (2009) for a helpful discussion of these approaches. While I think her approach is intuitively quite plausible, I don't think we have to accept a non-comparative account to deal with this particular objection.

because if you hadn't done this, somebody else would have. Thus, someone who accepts this account can only say that you harmed the lobster if they were to say that it's possible for you to harm the lobster when the lobster doesn't suffer any harm at all. It would be best not to say that.

Comparative accounts of harm tell us that a victim is harmed iff something makes the victim worse off.[14] Victims cannot be worse off *simpliciter*. Something harms a victim only if it makes the victim worse off than the victim is made to be under some relevant alternative. It's quite natural to think that the relevant alternative to consider in determining whether a happening harms someone is the one in which the happening doesn't happen because when we are trying to decide what to try to avoid or prevent, we run (if we're rational) a kind of counterfactual comparison test. The lobster has little reason to do what it can to get you to act otherwise if she knows that another patron will order her for dinner if you don't. The important point here is that it would be strange for the lobster to think that she won't be harmed by any of the patrons in the restaurant because so many of them want her boiled alive.

Consider an alternative to the counterfactual comparison account. The revised comparative account of harm tells us to compare an occurrence to its non-occurrence to determine whether the occurrence harmed the victim rather than comparing what happened to what would have happened if the occurrence hadn't occurred. Some events are intrinsically bad for a victim, cause something that's intrinsically bad for a victim, or prevent the victim from enjoying something that's intrinsically good. [15] An event harms a victim iff it meets this disjunctive condition and the occurrence (along with its effects) are worse for the victim than the non-occurrence's effects would have been. This account is tailored to vindicate the intuition that even if there are multiple patrons who will order the lobster to be boiled alive, the particular patron that calls in the order is the one that harms the lobster. The order initiates a causal chain that results in the lobster being boiled alive and then dying. This is a worse state than the state that the non-occurrence of the order would cause. The non-occurrence wouldn't cause the lobster to be in any state at all. This *isn't* because non-occurrences don't cause. It seems plausible that

---

14. See Bradley (2009) for a defense. When 'iff' is used, it is a shortening of 'if and only if'.

15. This is Bradley's (2009) notion of a *prima facie* harm. A complete account of harm should say something about *preventions* whereas the account that I've offered above doesn't. We should compare the goods that an occurrence prevents to the goods that the non-occurrence would have prevented. That's unwieldy, however, so I've simplified the account above accordingly.

non-occurrences can cause something (e.g., the failure to water a plant or feed a pet can cause death). The point is that this particular non-occurrence didn't cause the lobster to boil or to die, and this allows us to attribute responsibility for the lobster's bad state to one particular order.[16]

I don't think that consequentialists will like this account of harm. It's hard to see how a consequentialist could think that the concept of harm as it's understood here could be a useful concept for the purposes of moral theory since it doesn't have much to do with the kinds of counterfactual assessments that allow us to determine which of the available alternatives will involve the greatest amount of aggregate well-being. I've suggested that morally conscientious agents should be concerned about the particular role they play in harming and that this requires us to think about causal relations between themselves and the moral patients they causally interact with. Consequentialists will likely say that it would be odd for the morally conscientious agent to be concerned with this. They would say that we ought to be concerned with aggregates, not our role in producing these aggregate levels.

One of Williams's (1988) examples might help to illustrate the consequentialist's concern. You are offered a choice. You can pick up a rifle and shoot one villager held by the general's men or do nothing. If you do nothing, the general's men will execute this villager and nineteen others. You might think (as Williams presumably didn't) that it would be perverse for you to think of your role in the death of a villager as the sort of thing that should enter your deliberation in this sort of case. I agree. It would be perverse to do so when it's so clear that concerns about aggregate well-being override concerns you might have about your role in the killing.

Having said that, I think we're on pretty good ground here. First, the view that I'd defend isn't a view on which considerations about the role you play in harming, say, override considerations about overall levels of utility. I'd say that there's a *pro tanto* reason for you to refrain from shooting a villager and that this reason can potentially be overridden if you have a way of shooting this villager so as to save nineteen lives. As this is just what the consequentialist wants us to say, this sort of case needn't cause us any serious concern. Second, we've already seen that our concern should be directed primarily toward patients

---

16. The modified comparative account of harm applies some ideas of Sartorio's (2005, MS) about causation and responsibility to deal with problems for comparative accounts of harm. To my knowledge, she hasn't tried to use her account of causation and difference-making to deal with problems that arise for the counterfactual comparison account, but it seems like a natural application. If it's helpful, credit clearly goes to her. If it isn't, credit will have to fall to me. For a critical discussion of Sartorio's approach, see Driver (2007).

and their well-being and so only indirectly concerned with aggregates. If the consequentialist case against the kind of harm principle I'm offering rests on the idea that our concern as morally conscientious agents should be with aggregate levels of well-being, the force the objection derives from an approach we've already seen to be problematic insofar as it lacks the resources to explain why we should draw a moral distinction between promoting the interests of actual patients and bringing patients into being so as to promote their interests. Third, it would seem that the right way to test the account on offer is to consider a modified version of Williams's case. If the choice we're offered is one in which we can decide to shoot or refrain from shooting and that same villager will be shot by the general's men, I'd say that there's a duty to refrain. If you share that intuition, it counts in favor of the view on offer.

Finally, I should note that the account of harm offered here doesn't vindicate the idea that when you harm some victim, the victim has to be worse off overall. Some see this as a problem because they think that *to be harmed*, one has to be worse off overall as a result and think that if you harm someone, someone has to be harmed.[17] While I agree that if you harm someone, someone has to be harmed, I don't think that to be harmed, you have to be worse off overall as a result. Consider Ross's principle of non-maleficence, a principle that says that there's a *prima facie* duty not to harm. If we said that we had only a *prima facie* duty not to perform actions that leave someone worse off overall, we'd lose sight of the fact that any of our actions that cause something intrinsically bad for a moral patient calls for a justification and is *prima facie* wrong. It's true that considerations about overall well-being *might* provide that justification, but that's consistent with the idea that every harm (as I've characterized it) might be *prima facie* wrong to bring about. Thus, in moral contexts, at least, it makes sense to characterize harm as I have, provided that we understand that it's one moral notion among many and shouldn't be considered to the exclusion of considerations about whether a victim might be better off overall by being harmed than not. I suspect that part of the intuition that harm should have to do with overall well-being has to do with the fact that in many non-moral contexts where we're thinking about which option would be best for us, we rightly focus on aggregate well-being rather than particular intrinsically good or bad events. I'd say that our focus in these contexts *should* be on overall notions and that this just goes to show that the notion of harm isn't nearly as useful in such contexts as the distinct notion of what's better for us overall.

17. Thanks to Ben Bramble for raising a version of this objection.

A Final No-Difference Argument

A final challenge remains. Consider one final no-difference argument:

> *I can see that I might bear a special kind of responsibility for the causal con-*
> *tribution that I make, but the situation I face in a grocery store is unlike the*
> *preemption cases because my decisions in this setting don't make a causal dif-*
> *ference to the well-being of any animal. The animals I buy were already dead*
> *when I arrived at the store and the decision to slaughter more animals for*
> *food isn't sensitive to my decision about whether to buy chicken, lamb, beef,*
> *and so on. Since my actions don't have any causal impact on any living ani-*
> *mals, I don't have to take account of considerations of animal welfare in de-*
> *ciding what to buy in the store.*

Does this argument get the carnivore off the hook?

These arguments concern cases that might appear to be different in certain respects. The second no-difference argument focuses on preemption cases in which a particular patron's decision has the potential to cause the death of a moral patient. In this case, the consumer is trying to decide whether to purchase an animal that has already been killed. Moreover, the patron in the restaurant has very good reason to think that her decision can cause the death of the lobster. The consumer in the shop has very good reason to think that her decision won't harm any animal at all. She might say the following:

> [I]f I did not buy and consume factory-raised meat, no animals would
> be spared lives of misery. Agribusiness is much too large to respond to
> the behavior of one consumer. Therefore I cannot prevent the suffering
> of any animals. I may well regret the suffering inflicted on animals for
> the sake of human enjoyment. I may even agree that the human enjoy-
> ment doesn't justify the suffering. However, since the animals will
> suffer no matter what I do, I may as well enjoy the taste of their flesh
> (Norcross 2004: 231).

Isn't she right?

This is the infamous causal impotence defense. There's a familiar consequentialist response to this argument, one that strikes me as being decisive.[18]

---

18. I believe this response first appears in Almeida and Bernstein (2000). For further discussion of consequentialist responses to the threshold argument, see Almassi (2011), Chartier
(2006), Harris and Galvin (2012), Matheny (2002), Norcross (2004), and Kagan (2011).

It's true that agribusiness is too large to respond to the behavior of just one consumer, but it has to respond to some consumer behavior. Let's say that every month the factory gets data about the number of chickens sold and will change its production in response to increases or decreases in demand. If, say, demand decreases by one thousand, then one thousand fewer chickens will be produced. Of course, the consumer might have good reason to think that her decision not to purchase won't result in a reduction of the number of chickens killed, but why should that matter? If you think that the permissibility of an action is determined by *expected* utility rather than actual utility, this is irrelevant. It's been conceded that the utility gained by killing and eating a chicken is less than the amount of disutility involved in killing the chicken. Once that's conceded, it also has to be conceded that the expected utility of buying chicken under the conditions described is also negative. There is one chance in a thousand that the consumer purchases the threshold chicken. If purchased, it triggers the death of one thousand chickens, which is a very bad result. If not, it avoids triggering the death of one thousand chickens, which is a much less bad result. And in terms of expected utility, the expected utility of taking the chance is the same as the expected utility of raising a chicken in your own little factor farm and killing it yourself.[19]

Of course, the consumer might think to herself that it really doesn't matter whether *she* buys the threshold chicken or not because somebody is going to buy it. Perhaps, but then this consumer is relying on the arms dealer defense considered above. If it doesn't work for war criminals, it shouldn't work for unreflective carnivores. Imagine we gather all the consumers who purchased chicken in a particular grocery store and arrange them in line on the basis of their purchase. Every consumer, let's say, buys just one chicken. Suppose you are the seventh person in this line. You look back and see the consumer who purchased the one thousandth chicken. She triggered a slaughter of one thousand more chickens, but she wouldn't have if you didn't do your part. This was a team effort. You could have chosen not to make your own contribution, which causally is no different from hers, but you didn't and so you share equal causal responsibility for the deaths just as you would if the whole lot of you grabbed a rope and pulled a very heavy cage filled with these chickens over the side of a cliff.

Suppose, however, that you look back down the line and the people after you and it just so happens that nobody bought the threshold chicken. As a

---

19. For arguments that overall obligation depends upon expected value rather than actual value, see Jackson (1991) and Zimmerman (2008).

matter of luck you aren't part of a unit that worked together to kill off one thousand chickens. Maybe the factory closed its shutters or the grocery store burned to the ground. If you think you are under an obligation not to impose risk, this just shows that your actions didn't harm, not that you didn't commit a wrong. Compare this case to the case where you play a round of Russian roulette and live to tell the tale. No harm came of what you did, but you shouldn't have done it. Moreover, let's not forget that the scope of your obligation concerning animals can be quite broad. Your financial support sustains the agribusiness outfits that produced your chicken. Even if your previous supplier closes shop, your financial support sends signals to other producers who compete for your business. If we focus on just the factories that you buy from, we lose sight of this.

We also shouldn't lose sight of the fact that your behavior as a consumer helps to perpetuate the idea that it's acceptable for people to treat moral patients as resources to be used for making a profit with little regard to their welfare. Their behavior can be modified by public sentiment, but your buying behavior doesn't just encourage the further slaughter of animals, it helps to shield them from the moral judgments that could be efficacious in reforming their practices if only more people like you saw the light. Here's a conjecture: if a critical mass of people came to believe that it's morally repugnant for people to profit from an industry that causes widespread animal suffering, there would be fewer people who felt free to invest in it. If there were a social stigma attached to factory farming that's akin to the social stigma that's attached to, say, pornography or the exploitation of sweatshop labor, there would be fewer factory farms. Although you cannot choose whether there will be a critical mass of condemnation that would influence the behavior of investors, you can directly determine whether you'll be part of the critical mass that sustains a morally repugnant practice. As we've seen earlier, this is sufficient to show that you're under some obligation to refrain because it means that you're a partial cause of a state that impedes a triggering event that results in a significant decrease in the amount of animal suffering.

Even if you are reasonably convinced that your own actions won't help to alleviate animal suffering, there's one further consideration to bear in mind. The ground of your obligation to refrain from eating animals has to do with animal suffering. The scope of your obligation can be quite broad. It needn't be limited to an obligation to take steps to reduce animal suffering. It might require you to perform actions that have no causal connection to the welfare of any animals. If there are institutions or practices responsible for widespread suffering, you might be obligated to break all association with them even if doing so has no real causal impact on their operations. As social creatures, having a good social standing, one that's free from sanction and blame, might itself be a good, a good that's not appropriately enjoyed by those responsible for widespread death and suffering. If there were institutions responsible for the widespread death or the widespread abuse of children, for example, you wouldn't think that it would be acceptable for you to continue with any sort of involvement with these institutions by investing in them, working for them, or purchasing their products. The fact that we're comfortable sustaining connections with the institutions responsible for the widespread slaughter of moral patients suggests that we haven't yet internalized the obvious consequences of the marginal cases argument.[20]

References

Almassi, B. 2011. The Consequences of Individual Consumption: A Defense of Threshold Arguments for Vegetarianism and Consumer Ethics. *Journal of Applied Philosophy* 28: 396–411.

Almeida, M., and Bernstein, M. 2000. Opportunistic Carnivorism. *Journal of Applied Philosophy* 17: 205–211.

Arpaly, N. 2004. *Unprincipled Virtue*. Oxford University Press.

Bradley, B. 2009. *Well-Being and Death*. Oxford University Press.

____. 2012. Doing Away with Harm. *Philosophy and Phenomenological Research* 85: 390–412.

Carlson, E. 1995. *Consequentialism Reconsidered*. Kluwer.

Chartier, G. 2006. The Threshold Argument against Consumer Meat Purchases. *Journal of Social Philosophy* 37: 233–249.

Dancy, J. 1998. Wiggins and Ross. *Utilitas* 10: 281–285.

Driver, J. 2007. Attributions of Causation and Moral Responsibility. In W. Sinnott-Armstrong (ed.), *Moral Psychology*, Vol. II. MIT University Press, pp. 423–439.

Gardner, J. 2007. *Offences and Defenses*. Oxford University Press.

Harman, E. 2009. Harming as Causing Harm. In M. Roberts and D. Wasserman (ed.), *Harming Future Persons*. Springer, pp. 137–154.

____. 2011. Does Moral Ignorance Exculpate? *Ratio* 24: 443–468.

Harris, J., and R. Galvin. 2012. "Pass the Cocoamone, Please": Causal Impotence, Opportunistic Vegetarianism, and Act-Utilitarianism. *Ethics, Policy, & Environment* 15: 368–383.

---

20. I want to thank Ben Bramble, Bob Fischer, and an anonymous referee for their helpful comments on an earlier draft of this chapter. Thanks also to Matt Burstein, John Harris, Robert Howell, Alastair Norcross, and Amy Revier for discussing these issues with me.

Jackson, Frank. 1991. Decision-Theoretic Consequentialism and the Nearest and Dearest Objection. *Ethics* 101: 461–482.

Kagan, S. 2011. Do I Make a Difference? *Philosophy and Public Affairs* 39: 105–141.

Kamm, F. 2005. Ethical Issues in Using and Not Using Embryonic Stem Cells. *Stem Cell Reviews* 1: 325–330.

Kazez, J. 2010. *Animalkind: What We Owe to Animals*. Blackwell-Wiley.

Littlejohn, C. 2014. The Unity of Reason. In C. Littlejohn and J. Turri (ed.), *Epistemic Norms*. Oxford University Press, pp. 135–154.

Marquis, D. 1989. Why Abortion Is Immoral. *Journal of Philosophy* 86: 183–202.

Matheny, G. 2002. Expected Utility, Contributory Causation, and Vegetarianism. *Journal of Applied Philosophy* 19: 293–297.

Norcross, A. 2004. Puppies, Pigs, and People: Eating Meat and Marginal Cases. *Philosophical Perspectives* 18: 229–245.

Ross, D. 1930. *The Right and the Good*. Oxford University Press.

Sartorio, C. 2005. Causes as Difference Makers. *Philosophical Studies* 123: 71–96.

_____. 2015. What Difference Does it Make? On Acting Freely and Making a Difference. Unpublished Manuscript.

Singer, P. 1993. *Practical Ethics*. Cambridge University Press.

Thomson, J. 1971. A Defense of Abortion. *Philosophy and Public Affairs* 1: 67–95.

Williams, B. 1988. Consequentialism and Integrity. In S. Scheffler (ed.), *Consequentialism and its Critics*. Oxford University Press, pp. 20–50.

Zimmerman, M. 2008. *Living with Uncertainty*. Cambridge University Press.

7 A MOOREAN DEFENSE OF THE OMNIVORE

Tristram McPherson

Introduction

Most of us are unreflective omnivores.* We consume meat and other products made from or by non-human animals without pausing to wonder whether doing so might be wrong. The permissibility of this behavior can seem like a bit of ethical common sense. However, common sense can be challenged, and the acceptability of eating meat has been challenged repeatedly. Sometimes the challenge takes the form of grisly images of cruelty to animals, or blunt slogans like "Meat is Murder." However, the challenge also takes another, more reasoned, form. Many philosophers have argued forcefully that eating meat is wrong. A conscientious omnivore might seek to wade through this sea of arguments and try to find flaws in each one, but that would be a Herculean task. This chapter examines another way of trying to answer these ethical challenges to the omnivore's lifestyle. This is to adapt a form of argument originally developed by G. E. Moore to argue against the skeptic (who

* I am indebted to an audience at Charles Sturt University for vigorous and helpful discussion of ideas that went into this chapter. I am also indebted to Derek Baker, Mark Budolfson, David Faraci, Bob Fischer, and David Plunkett for illuminating comments. This paper chapter draws in places on ideas previously developed in my (2012), (2014), and especially (2009).

claims that we know vastly less than we ordinarily think we do) and other revisionary philosophical views. Because Moore used his argument to defend commonsensical views about knowledge and reality against revisionary challenges, the omnivore might hope that Moore's style of argument would be similarly effective against the revisionism of the ethical vegetarian.

This chapter examines the attempt to adapt Moore's argument on behalf of the omnivore. I first introduce and explain Moore's argument against the skeptic. I then explain how that argument can be adapted to address two influential philosophical arguments against the omnivore, due to Tom Regan and James Rachels. The adapted Moorean arguments appear strikingly similar to the original. However, I go on to argue that we should not simply assume that all Moorean arguments are created equal. Instead, I propose a set of principled criteria that can be used to test Moorean arguments on a case-by-case basis. Those criteria give the Moorean reason for optimism against the skeptic, but suggest that the Moorean's case is much weaker against the ethical vegetarian. I conclude that the Moorean omnivore's argument has potentially uncomfortable implications for all sides in debates about ethical vegetarianism, and illuminates important and neglected questions about the force of philosophical arguments in applied ethics.

A Brief Introduction to Moorean Arguments

In order to explain the Moorean strategy, it will be useful to have a skeptical argument as a foil. Arguments for skepticism are many and varied, but I will focus on a single simple skeptical argument:

1. If my evidence would appear identical given a certain hypothesis, then I do not know that hypothesis to be false
2. My evidence would appear identical if I were a brain in a vat whose (overwhelmingly false) beliefs about the world were responses to the ways that the scientists who run the vat stimulated my brain, rather than to ordinary sensory contact with the world
3. If I do not know that I am not a brain in a vat, then (among *many* other things) I do not know that I have hands

C. I do not know that I have hands

This argument talks about *evidence*: think of evidence as the set of reasons someone has for thinking that a claim is true. This argument is represented as a series of numbered premises. That helps to distinguish each of the key claims that the skeptic appeals to in supporting her conclusion. All of the premises of this argument are needed to support the conclusion: The first two premises entail that you do not know that you are not a brain in a vat. The third premise claims that if this is true, then you do not know many seemingly obvious propositions about the world, such as the proposition that you have hands.

What is the anti-skeptic to do, when confronted with this argument? The argument can be seen as implicitly posing a challenge to the anti-skeptic: "So you don't like my conclusion? Well, if each of these three claims are true, that conclusion *must* be true. So tell me what you think is wrong with one of these claims!" Understood as posing a challenge in this way, the argument appears powerful, because each of its premises seems very credible. I will briefly explain why each of these premises is hard to resist.

The first premise can be motivated by example. Suppose that you have friends with identical twin toddlers Allie and Sally. One day you visit them, and your friends have dressed their daughters identically. While this is adorable, you find that it is impossible for you to distinguish the twins when they are dressed this way. Suppose that one of them runs by, and you form the belief that it was Allie. Because you cannot tell the twins apart, your visual evidence would have appeared identical if it had been Sally who ran by. This seems to show that you do not in fact know that it was Allie and not Sally who ran by. The first premise of the skeptic's argument seems to be an excellent explanation of why this is so.

Next consider the second premise. It seems imaginable that sophisticated scientists could keep someone's brain alive in a vat, and immerse that person in a virtual reality by directly stimulating relevant parts of the brain. If the scientists were sophisticated enough, it further seems possible that the evidence the person received as a result of such stimulation would be indistinguishable by that person from an ordinary sensory experience. After all, the brain is connected to the sense organs and the rest of the body by a complex series of electrochemical connections, so it seems that it should in principle be possible to mimic those connections. This means that the hypothesis described by the second premise appears possible, making the second premise hard to deny.

The third premise also appears very plausible. In order to know that you have hands, it must be true that you do have hands. But brains in vats don't have hands. So (by the same reasoning that supported the first premise) you seemingly cannot know that you have hands without knowing that you are not a brain in a vat.

To be clear, philosophers have found important ways to challenge each of these premises, but doing so is clearly a difficult task. The Moorean responds differently, rejecting the skeptic's conclusion *without* explaining what is wrong with any of the skeptic's premises. By doing so, the Moorean refuses to take up the skeptic's challenge, mentioned above. Because of this, some have taken the Moorean to be objectionably dogmatic. The Moorean, on this reading, fails to "follow the argument where it leads."

A much more charitable understanding of the Moorean is possible, as I will now show. Imagine combining the skeptic's three premises into one big claim: call this the *conjunction* of the skeptic's claims. It is agreed on all sides that the skeptic's conjunction is inconsistent with *I know I have hands*. Suppose that you notice that you accept the conjunction, that you also accept *I know I have hands*, and that this means that you accept inconsistent claims. One way to address this inconsistency is to recheck the apparent evidence you have for each of the inconsistent claims. For example, if such rechecking convinced you that you have no reason to accept the first premise of the skeptic's argument, this would allow you to resolve the inconsistency by rejecting that premise. But suppose that no such resolution is available. Then you should presumably attempt to *weigh* the evidence you have for each of the inconsistent claims against the evidence you have for the other.

The important thing to notice, from the Moorean's point of view, is that you can have much stronger evidence for a claim (such as *I know I have hands*) than for a conflicting claim (such as the skeptic's conjunction) without knowing why the conflicting claim is false. Consider an analogy: suppose that after dinner at the restaurant, your four friends each independently calculate the split bill, while you savor your dessert. Three of your friends conclude that each person owes exactly $18.62, while the fourth concludes that each owes $16.27. Other things being equal, this would give you good reason to believe that your three friends got it right, because there is little chance that they all miscalculated in precisely the same way. But notice that this reasoning does not (and need not) shed light on *why* your fourth friend's conflicting answer is wrong. If the analogy holds, the skeptic's challenge may be illegitimate: it may be sensible to reject the skeptical argument without explaining where it goes wrong.

The Moorean embraces this idea, claiming that the evidence for *I know I have hands* is much stronger than the evidence for the skeptic's conjunction, and that it is thus reasonable to continue to believe one has hands, even in face of the conflict with that conjunction.[1] It is worth emphasizing that for the

---

1. Compare Moore (1959, 226), where he replies in roughly this way to an argument due to Bertrand Russell.

Moorean to be correct about this, the evidence for *I know I have hands* needs to be significantly stronger than the evidence for the conjunction, and not merely slightly stronger. This is because if your evidence for each of two conflicting claims is close to equally strong, you should suspend judgment about both of them. To see this, suppose that you have been looking at apartments to rent with your friends, Chuck and Diane. You try to remember the color of the kitchen in the second apartment, but draw a blank. When you ask your friends, Chuck is confident that the kitchen was beige, and Diane that it was yellow. Suppose that you have reason to trust Diane a bit more on these things; she is usually slightly more attentive to visual details than Chuck. So, if you needed to bet on the color of the kitchen, it would make sense to bet it was yellow. Still, given your friends' conflicting testimony, you should not believe either that the kitchen was beige or that it was yellow, until you get further evidence: you should suspend judgment about both claims.

The Moorean Case for the Omnivore

With the Moorean's basic reasoning in hand, I will now explain how this reasoning can be adapted by the omnivore. The case against the skeptic used the skeptic's argument as a foil, so I begin by introducing foils here as well. As with skepticism, philosophical arguments for vegetarianism are many and varied. One important way that they vary is in the *scope* of the ethical principles that the vegetarian defends and applies. Some philosophical vegetarians develop and then apply comprehensive ethical theories. Others appeal to local ethical principles that do not purport to explain the rightness or wrongness of every action. Consider an influential instance of each approach.

Tom Regan's case for animal rights (2004) develops and applies a comprehensive ethical theory. Regan argues that individuals possess various moral rights, which directly reflect the inherent moral worth of those individuals. By seeking to ground rights directly in moral worth, Regan raises a pressing question. On any plausible view of rights, some things (e.g., you and I) possess rights (and hence inherent moral worth), while others (e.g., a shard of broken plastic) do not. What explains the difference? Regan argues that many initially plausible answers to this question are indefensible. For example, consider the idea that inherent moral worth requires capacities for ethical agency or sophisticated rational thought. This would entail that non-human animals lack rights. However, it would also entail that many humans—for example young children and severely mentally handicapped adults—lack rights. And this is implausible. Or consider the idea that having moral worth requires being a member of the species *homo sapiens*. This avoids the problems facing the rational capacity idea, but it looks like an attempt to explain a fundamental ethical property by appeal to something ethically irrelevant. To see this, imagine that we discovered an alien species with capacities to think, feel, love, and act that are very like our own. Mere difference in their genetic code surely cannot deprive them of rights. According to Regan, the only defensible alternative is that a sufficient criterion for having intrinsic worth is being the experiencing subject of a life (2004, §7.5). Since many of the animals that humans eat and otherwise use are experiencing subjects of lives, Regan concludes that these animals have moral rights that are just as strong as ours.

Arguments for vegetarianism need not appeal to ambitious claims about the structure of morality, as Regan's does. James Rachels's (1977) argument begins with the simple and plausible thought that it is wrong to cause suffering and death to sentient beings unless one has a strong reason to do so. Contemporary ways of raising animals for food cause an extraordinary amount of suffering and death to sentient beings, and Rachels argues that the various reasons that humans typically have for raising animals for meat do not constitute a strong enough reason to outweigh the presumptive wrongness of this practice. Contemporary practices of raising animals for meat are thus wrong. It might be noted that the ordinary omnivore does not *cause* animals to suffer or die: she buys her meat already dead. In order to address this point, Rachels (like Regan) argues for the further thesis that it is wrong to purchase and consume meat, given how ethically objectionable its production is.

These arguments do not merely imply that we should stop eating meat. The suffering that Rachels objects to is a feature of most contemporary animal farming, so his argument would generalize to rule out eating eggs or dairy products. Further, the amount of suffering inflicted by contemporary animal farming boggles the mind: many *billions* of animals suffer and die every year in the system that produces our meat. Following James Rachels in admitting the ethical significance of animal suffering thus appears to suggest that our contemporary relationship to farm animals is not merely wrongful; it is an ethical catastrophe.[2] Regan's view arguably has even more radical implications: it makes the omnivore's dinner and the cannibal's appear strikingly ethically similar. These arguments raise further important questions about whether the ethical vegetarian should even tolerate the company of omnivores. (For a vivid depiction of someone struggling with the radical

---

2. Stuart Rachels (2011, especially §6) makes this case forcefully.

implications of related ideas, see J. M. Coetzee 2001, especially 19–20 and 68–69.) In light of these considerations, it is fair to describe these leading philosophical arguments for vegetarianism as radically revisionary of common ways of thinking about the ethics of our relationship to animals.

This description encourages the thought that the omnivore can cast herself in Moorean clothes, as the defender of common sense against philosophical attack. Theses such as *it is not wrong to drink a glass of milk* or *it is not wrong to eat some chicken* appear to be bits of ethical common sense: they are widely held, and typically taken for granted as obvious. (Of course, if you are reading this volume, you may *not* take these claims for granted. And this may lead you to doubt that they really are commonsensical. But the same would seemingly be true of *I know that I have hands*, if you were earnestly reading a volume on skepticism.) The thought that the omnivore is warranted in adopting the Moorean guise is further encouraged by Kit Fine's gloss on the core Moorean idea:

> [I]n this age of post-Moorean modesty, many of us are inclined to doubt that philosophy is in possession of arguments that might genuinely serve to undermine what we ordinarily believe (2001, 2).

Fine's suggestion here amounts to the idea that—systematically—commonsense beliefs are more reasonable than the conclusions of any philosophical arguments inconsistent with them.

The structure of the Moorean reply to our two philosophical arguments for vegetarianism is the same as it was against the skeptic. First, join together all of the claims required by the vegetarian to make her argument. Then note that together, these claims are inconsistent with the truth of a commonsensical claim like *it is not wrong to drink a glass of milk*. Then claim that it is most reasonable, in the face of this conflict, to retain the commonsensical claim.

Fine's suggestion provides reason for optimism about this strategy. After all, Regan's and Rachels's arguments are both uncontroversially *philosophical*. Thus, Regan's argument requires the premises that *being an experiencing subject suffices for possessing rights* and that *it is wrong to consume products that were produced in ways that violate rights*, both of which are substantial theoretical claims defended by philosophical argument. Similarly, Rachels's argument requires the premises that *it is wrong to cause suffering (even to animals) without very good reason*, that *there are insufficient reasons to justify causing the suffering involved in the production of meat and dairy*, and that *it is wrong to consume products that were produced wrongfully,* all of which are significant ethical claims defended by philosophical argument.

These examples illustrate a general recipe that the Moorean omnivore can use against arguments for vegetarianism. Given the broadly commonsensical status of omnivorism, it is plausible that any argument against the omnivore will need to appeal to some philosophically substantive ethical claims. These might be ambitious theoretical principles (like Regan's rights view), or more local ethical principles (like Rachels's). Or they might be at some intermediate level of generality. But in any such case, a relevant inconsistent package can then be constructed. And then Moorean modesty about the force of philosophy against common sense will suggest that the omnivore's thesis should be retained in such conflicts.

Evaluating the Moorean Omnivore

So far, I have sought to explain the Moorean omnivore's strategy, and present it in a sympathetic light. I now turn to the task of evaluating that strategy. In order to do so, I will propose and apply a general set of criteria for assessing claims about the relative strength of bodies of evidence. I begin with some critical comments about Kit Fine's suggestion, which motivate the search for such general criteria.

Fine suggests that philosophy lacks arguments that could give us reason to abandon what we ordinarily believe. We should ask: is this suggestion (if true) explained by the special strength of common sense or by the special weakness of philosophy? The latter is undoubtedly the more plausible version of the view. After all, common sense should yield in the face of well-confirmed scientific hypothesis. Consider for example the foolishness of trying to deploy common sense against the hypothesis (entailed by Einstein's theory of general relativity) that space itself is curved.

Fine's suggestion is thus best understood as espousing modesty about what philosophical arguments can lead us to reasonably believe. This modesty may seem both reasonable and appealing, given the long history of philosophers' succumbing to intellectual overreach. The appeal of this attitude collapses under scrutiny, however, when we consider its application to ethics. Adopting Finean modesty here would threaten to transform the field into an exercise in apologetics, limited to generating explanatory frameworks to underlie, organize, and extend the views that we all already accept.

Of course, the fact that modesty about the force of philosophical argument is unappealing does not show that it is unreasonable. But here again, reflection on ethics suggests reasons for doubt. Consider a historical context in which the ethical acceptability of radical racial and gender inequality was generally treated as commonsensical. Imagine someone bringing careful exposition of our best arguments against such inequality to ordinary people in this context. Set aside how persuasive such exposition would likely have been. Instead ask: could our speaker's audience be *reasonable* in holding unreformed racist or sexist ethical beliefs after understanding the speaker's arguments? I strongly suspect not. More modestly, Fine's suggestion is not obviously plausible when we consider such cases.

This suggests that we should not take the reasonableness of all Moorean arguments for granted. We should instead seek a principled method for evaluating them. Recall that, rather than challenging one of the premises of a revisionary argument, the Moorean claims that the evidence supporting her commonsensical thesis is much better than the evidence supporting the revisionist's premises. Because of this, evaluating a Moorean argument essentially involves assessing claims about the relative strength of the evidence that supports conflicting claims. Performing this evaluation therefore requires us to find *second-order* evidence: evidence concerning how strong our evidence for a claim is.

One might worry that there will be nothing principled to say here: after all, one might expect it to be at least as hard to assess second-order evidence as it is to assess the first-order evidence directly. The approach that I propose is tailored to meet this concern. The basic idea is to identify general features, the presence or absence of which would, other things being equal, reasonably lead us to raise or lower our estimation of how well-supported belief in a claim is. I call such features *generic indicators of strength of evidence*.[3] Because they are generic, the indicators I propose are well suited to provide principled (although defeasible) support for one view over another, in a controversy over the strength of evidence for a claim.

I cannot canvass every possible such indicator here. I focus on five indicators that I take to be crucial for evaluating Moorean arguments. These indicators are:

1. Relative confidence in the Moorean and revisionary theses
2. Prevalence of philosophically naïve proponents of the conflicting theses
3. Extent and nature of the change to our beliefs required by the revision
4. Relative fit of the conflicting theses with our epistemic paradigms
5. Relative vulnerability of the conflicting theses to debunking explanations

---

3. For further discussion of the generic indicators idea, see my (2009) and (2012).

A *Moorean thesis* is just the central claim that some Moorean hopes to defend. In our cases, the Moorean thesis is: *it is not wrong to drink a glass of milk*. A *revisionary thesis* is the claim targeted by a Moorean argument. In our cases, it is: *it is wrong to consume animal products*. I now briefly explain the significance of each of the five indicators mentioned.[4]

First, consider the idea that one's degree of confidence in a belief is second-order evidence of how strong one's evidence for that belief is. This idea is required by ordinary trust in oneself as a believer. To see this, imagine what one's inner life would be like, if one took one's relative confidence to be no guide to what to believe. I tenuously believe that I left my keys on the desk, but (looking at the desk) I find myself very confident that I do not see them, that my eyes are functioning normally, and that my keys have not become invisible. It would be absurd to discount this confidence when considering what to believe about the location of my keys. Like all of the indicators, the evidence provided by confidence can be undercut. I might learn from experience, for example, that I have systematically inflated confidence in some of my beliefs about my own ability or performance. This would give me reason to discount that confidence.

The second indicator is non-philosophical consensus. One case for the significance of consensus simply generalizes trust in oneself. The idea is that, if one trusts one's own thinking, and one has no evidence that others are worse thinkers on a topic (or deceitful), one should extend one's trust to those others. This appears especially plausible where others speak with one voice. This indicator's focus on *non-philosophical* consensus is more controversial: if we took philosophers to be experts on a question and non-philosophers to be amateurs, it would seemingly make more sense to consider the weight of philosophical opinion. (Compare the way that it will often be reasonable to follow scientific opinion in the face of non-scientific consensus.) The Moorean project is most promising if we assume that philosophers are *not* experts on the relevant topic. For if they were, the non-philosophical consensus that we know we have hands, for example, would have little significance against the skeptic.[5] This is a complicated issue, but I will grant the Moorean this assumption for the sake of argument.

The third indicator is the extent and nature of the change to our beliefs required by the revisionary thesis. The basic idea here is straightforward: we

---

4. I defend these indicators in more detail in my (2009).

5. For one argument against the idea that philosophers are moral experts, which could be generalized to other topics of philosophical controversy, see McGrath (2008, §5).

tend to have evidence for what we believe. As a revisionary thesis challenges more (and more topically diverse) of our beliefs, we thus have reason to believe that it conflicts with more and more of our evidence.

The fourth indicator concerns the relative fit of the Moorean and revisionary theses with our *epistemic paradigms*. These are our most central and clear guides to evidence and reasonable belief. These paradigms include sources of evidence (for example: sense experience), and ordinary inferential practices (for example, predicting that a past regularity will continue). They also include paradigms of epistemically successful enquiry (such as scientific enquiry). Together, these paradigms make up much of our competence to evaluate epistemic claims. It is plausible that broad inconsistency with such paradigms is a powerful indicator that something has gone wrong with a piece of reasoning.

The final indicator is the relative vulnerability of the Moorean and revisionary theses to *debunking explanations*. An explanation debunks a claim if it offers a compelling explanation for the claim's apparent evidential status that gives us good reason to reject that status. For example, the fact that *everything looks red in this light* debunks my visual evidence that the paper I am holding is red. A claim is more or less *vulnerable* to debunking explanation depending on how credible alleged debunking explanations of that claim are. If a claim is substantially vulnerable to debunking, this suggests that the evidence for it may not be nearly as strong as it otherwise appears to be.

With the five indicators briefly described and motivated, consider how they apply to the Moorean argument against the skeptic introduced above. Very briefly:

1. (*confidence*) I am enormously confident that I have hands. And the most radical skeptical doubts appear to have little persistent effect on this confidence. As David Hume noted, an evening of pleasant company and diversion is often enough to render skeptical speculation inert (2001 [1740] §1.4.7).
2. (*philosophically naïve acceptance*) Outside of the seminar room, radical skepticism is an extremely rare phenomenon, while belief in the possession of hands is ubiquitous, except in cases that are irrelevant to the skeptical challenge. (Some of us *do* lack hands and reasonably believe that.)
3. (*extent of revision required*) The skeptic's thesis requires wholesale abandonment of almost all of one's beliefs about almost all topics.
4. (*fit with epistemic paradigms*) The skeptic's thesis is inconsistent with many of our epistemic paradigms, since it would require us to distrust or abandon them.

5. (*vulnerability to debunking*) Independent of the skeptical challenge at issue, the belief that I have hands does not appear particularly vulnerable to debunking explanation (although the skeptic's revisionary argument is not particularly vulnerable in this respect either).

Together, these indicators constitute a significant case that the Moorean thesis is much better supported by our evidence than the skeptic's revisionary alternative.

I take the generic indicators account just sketched to provide the most promising principled defense of the strength of traditional Moorean arguments against the skeptic. The account focuses on the crucial question of how to assess competing claims about the relative strength of competing evidence. The account appeals to informative and seemingly relevant considerations. And the Moorean case shines relative to these considerations. This account also protects the Moorean from implausible overgeneralization: for example, it will not vindicate implausible Moorean attacks on well-supported but surprising scientific hypotheses.[6]

These virtues of the generic indicators account warrant treating that account as a criterion against which to assess more controversial Moorean arguments. I therefore now consider how the generic indicators apply to the Moorean omnivore's argument:

1. (*confidence*) Many of us are very confident that it is not wrong to drink a glass of milk, or eat a steak, even after encountering arguments for vegetarianism. However, the case here appears weaker than versus the skeptic. Most directly, it seems both rare and unreasonable to be as confident that it is okay to eat a steak, as one is that one has hands.
2. (*philosophically naïve acceptance*) The omnivore's key claims are widely accepted. However, we also see a small but significant group of non-philosophical vegetarians, living their commitment to the revisionary thesis. By contrast, it is controversial whether anyone can fully embrace and live the skeptical perspective.

   In both of these cases, the indicator arguably does something to bolster the credibility of the Moorean argument, but the indicators are not as favorable as they were against the skeptic. The omnivore cannot make even this modest claim about the remaining indicators.

---

6. See my (2009, 14–15) for discussion and defense of this claim.

3. (*extent of revision required*) Consistently adopting vegetarianism is clearly a significant change in one's ethical outlook. However, compared with skepticism, it involves *very* limited adjustment to one's beliefs. For example, it will leave untouched almost all of one's non-ethical beliefs and many of one's ethical beliefs. In light of this, the Moorean omnivore fares poorly relative to the third indicator.

4. (*fit with epistemic paradigms*) Accepting the vegetarian's revisionary argument poses no significant threat to our epistemic paradigms. (The Moorean might insist that trust in commonsense claims is itself an epistemic paradigm. If so, the vegetarian's argument threatens only this, as every revisionary argument must.) Indeed, the philosophical defender of vegetarianism might argue that, given the strength of arguments for her view, belief in vegetarianism might be an instance of being appropriately sensitive to those paradigms. So the Moorean omnivore fares poorly relative to the fourth indicator.

5. (*vulnerability to debunking*) The omnivore's Moorean thesis is substantially vulnerable to debunking explanation (as I now explain), and so the omnivore does poorly relative to the fifth indicator.

The most promising debunking argument against the Moorean omnivore's thesis (and related beliefs) is perhaps the suggestion that these are products of what I will call *status quo rationalization*. A belief is a product of status quo rationalization just in case that belief is maintained because it vindicates the goals and behaviors of the believer and of others that the believer identifies as members of her ethical community. Consider as an example a member of a slave-owning family in the antebellum South, for whom owning and using slaves is a deeply embedded part of everyday life, and the life of those he is closest to. It is easy to predict that other things being equal, such a person will tend not to believe that slavery is a moral monstrosity. This is because taking oneself and those one identifies most closely with to be doing something seriously wrong makes for a particularly uncomfortable form of cognitive dissonance: very few people can comfortably identify themselves and their loved ones as morally bad (for dramatization of this point, see Coetzee 2001, 68–69). The more deeply embedded a behavior is in one's life—the more convenient or beneficial or pleasant; the more unquestioned by one's peers and loved ones— the more likely it is that the cognitive dissonance will be resolved by one's *values* changing in order to rationalize one's behavior, and the behavior of those one identifies with. This is relevant because eating animal products is deeply embedded in most of our lives, in a way apt to produce such rationalization. A belief is especially *vulnerable* to such debunking explanation where we lack explanatorily satisfying arguments for its truth. A contrasting example: it would surely be very uncomfortable to abandon the belief that wanton killing is wrong; but we possess satisfying arguments that provide plausible explanations of why it is wrong, so this belief is not very vulnerable to this sort of debunking.[7]

To sum up: the Moorean omnivore can appeal to confidence and non-philosophical consensus as modest indicators of the strength of her argument. However, unlike the anti-skeptic, she seemingly cannot appeal to the objectionably broad scope of the revision in question or to the revision's inconsistency with our epistemic paradigms. Finally, the confidence and consensus that the Moorean omnivore appeals to are vulnerable to significant debunking explanation, in a way that the Moorean anti-skeptic appears not to be. Recall that Moorean arguments have a very high standard for success: they need to make it reasonable to *retain belief* in the face of a competing argument, rather than either accepting the competing argument or suspending judgment. The five indicators canvassed strongly suggest that it is not plausible that the omnivore meets this standard. In light of this, the Moorean strategy fails to show that it is reasonable to be an omnivore in the face of philosophical arguments for vegetarianism.

Conclusions and Implications

This chapter has explored a Moorean strategy for defending the omnivore's lifestyle against the challenge posed by philosophical arguments for ethical vegetarianism. This strategy initially appears attractive, because of the structural similarity of the Moorean omnivore's argument to the standard Moorean argument against the skeptic. However, I have argued that once we develop the tools to offer a principled assessment of possible Moorean arguments, we find that the omnivore's argument falls well short of the standard set by the case against the skeptic. I take my argument to be too brisk to be utterly decisive: it is best understood as providing a strong initial challenge to the reasonableness of being an omnivore on Moorean grounds. This challenge invites the omnivore to look elsewhere for a reasoned defense of her views: perhaps she can appeal in some other way to the apparently commonsensical status of omnivorism, or to piecemeal diagnoses of what goes wrong

---

7. For more discussion of this debunking argument, see my (2014).

with the many ethical arguments for vegetarianism, or to a novel positive ethical argument for omnivorism.

In addition to this main challenge, my exploration of the Moorean omnivore's argument has interesting implications for the ambitions of philosophers, for Mooreans, and for the dialectic about vegetarianism. I close by briefly exploring these issues in turn.

First, my argument suggests that not all "common sense" is created equal. In light of this, we should reject Kit Fine's proposal that common sense is quite generally immune to philosophical revision. Further, the five indicators discussed here can be seen as providing a (provisional and sketchy) outline of the conditions under which philosophers can hope to reasonably challenge common sense. This is of crucial interest for philosophers, because the philosophical project looks very different depending on whether common sense is a legitimate target. Where common sense reigns supreme, philosophers are limited to helping us to better understand the philosophical alternatives that are compatible with common sense. Where common sense does not reign supreme, philosophers may reasonably retain hopes of leading us closer to the truth—and away from the grip of rationalization and ideology—on matters of importance. This chapter suggests that philosophers may retain such hopes concerning the ethical case for vegetarianism, and it is natural to think that the argument of this chapter would generalize at least to other pressing issues in applied ethics.

Second, the failure of the Moorean omnivore's argument is crucially a matter of *degree*. If I am correct, her argument falls well short of the standard set by the paradigmatic Moorean arguments and also well short of the standard required to secure knowledge. This does not mean that Moorean considerations have no probative force against the vegetarian, however. It is a difficult matter to assess exactly how much force they have. It *might* be argued on the basis of the first two indicators that the omnivore's Moorean considerations have just enough force to prevent arguments for vegetarianism from succeeding: that is, reflection on the Moorean argument might require us to suspend judgment concerning whether it is wrong to eat meat.

To be clear, I do not think that the Moorean argument succeeds even this much, but adequately defending this claim would require considerable (and difficult) argument that I have not provided. And this possible outcome of the debate should be of interest both to philosophers attracted to Moorean arguments and to anyone interested in the ethical debate between the omnivore and the vegetarian.

The idea that a Moorean argument should lead us to suspend judgment is an important issue for Moorean philosophers to grapple with. It is also uncomfortable: Moore was in the business of trying to preserve knowledge, not destroy it. But once we notice that candidate Moorean arguments will fall on a spectrum, with Moore's own paradigms at the strong end, it seems inevitable that at least some such arguments will entail that we ought to suspend judgment.

The possibility that the philosophical arguments for vegetarianism and the Moorean case for the omnivore together deprive us of knowledge concerning what we may ethically eat is also practically important. It would suggest that the proponent of ethical vegetarianism and of omnivorism each believe more than is reasonable. This would be unnerving on all sides. It also raises the question of what we should do, if faced with this moral ignorance.

This is itself a vexed philosophical issue.[8] I am inclined to think that a precautionary principle may apply here: other things being equal, we have reason to avoid doing things that we have strong (even if not decisive) reasons to believe are wrong. Suppose this is right, and that, at that end of the day, while the philosophical vegetarian's case gives us reason to believe that it is wrong to eat meat, it falls short of securing knowledge. In this case the precautionary principle would suggest that the vegetarian's argument should govern our behavior, even if it does not command our beliefs.

References

Coetzee, J. M. *The Lives of Animals*. Princeton: Princeton University Press, 2001.

Fine, Kit. "The Question of Realism." *Philosophers' Imprint* 1.1 (June 2001). 1–30.

Hume, David. *Treatise of Human Nature*. Eds. D. F. Norton and M. J. Norton. Oxford: Clarendon, 2001.

McGrath, Sarah. "Moral Disagreement and Moral Expertise." *Oxford Studies in Metaethics* Vol. 3. Ed. Russ Shafer-Landau. Oxford: Oxford University Press, 2008. 87–107.

McPherson, Tristram. "A Case for Ethical Veganism: Intuitive and Methodological Considerations." *Journal of Moral Philosophy* 11.6 (2014). 677–703.

———. "Moorean Arguments and Moral Revisionism." *Journal of Ethics and Social Philosophy* 3.1 (June 2009). 1–25.

———. "Unifying Moral Methodology." *Pacific Philosophical Quarterly* 93 (2012). 523–549.

---

8. For a view on this question broadly sympathetic to my proposal below, see Moller (2011); for a countervailing view, see Weatherson (2014).

Moller, Dan. "Abortion and Moral Risk." *Philosophy* 86.3 (2011). 425–443.

Moore, G. E. 1959. *Philosophical Papers*. London: George Allen & Unwin.

Rachels, James. "The Moral Argument for Vegetarianism." *Can Ethics Provide Answers? And Other Essays in Moral Philosophy*. Lanham, MD: Rowman and Littlefield, 1997. 99–107.

Rachels, Stuart. "Vegetarianism." *Oxford Handbook of Animal Ethics*. Eds. T. Beauchamp and R. G. Frey. Oxford: Oxford University Press, 2011. 877–905.

Regan, Tom. *The Case for Animal Rights*. (2nd ed.) Berkeley: University of California Press, 2004.

Weatherson, Brian. "Running Risks Morally." *Philosophical Studies* 167.1 (2014). 141–163.

8 THE CASE AGAINST MEAT

Ben Bramble

Introduction

There is a simple but powerful argument against the human practice of raising and killing animals for food (RKF for short). It goes like this:

1. RKF is extremely bad for animals.
2. RKF is only trivially good for human beings.
   So,
3. RKF should be stopped.[1]

Call this The Case Against Meat. Many consider The Case Against Meat to be decisive. But not everyone is convinced by it. Four main objections have been proposed:

1. *The first premise is false.* RKF is not extremely bad for animals, or at least, given the possibility of free-range farming, it needn't be. In fact, by giving animals an existence, RKF may even be in the best interests of these animals.

---

1. The classic statement of this style of argument is from Singer (1975).

2. *The second premise is false.* RKF is far more than merely trivially good for human beings. This is because of the pleasures of eating meat and what these contribute to various social and cultural aspects of our lives.
3. *Animal welfare is relatively unimportant.* Even if both premises of the argument are true, animal welfare is nowhere near as valuable as human welfare. It simply doesn't matter as much how they fare.
4. *As individuals, we are powerless to change anything.* Even if it would be best if RKF were to stop, none of us has any reason to abstain from eating meat. This is because our individual purchasing decisions have only a negligible effect on the demand for meat, and so none at all on RKF.

In this chapter, I will attempt to shore up The Case Against Meat by providing new responses to each of these objections.

The First Premise

Many have claimed that RKF is good for animals by giving these animals an existence. Leslie Stephen, for example, writes:

> The pig has a stronger interest than anyone in the demand for bacon. If all the world were Jewish, there would be no pigs at all.[2]

But it is implausible that having an existence can be better for a being than having no existence at all. To be better off in one scenario than in another one must have a level of well-being in *both* scenarios. But those who do not exist in a given scenario are not poorly off in that scenario. Rather, they have *no* level of well-being in that scenario.

It may be objected that non-existent beings *can* be well or poorly off. Cinderella, for example, does not exist, but she was very poorly off until she met her Fairy Godmother.

But when we talk about Cinderella, we are not saying "there is some woman who had evil stepsisters, rode to a ball in a pumpkin, fell in love with a prince, had a level of welfare," and so on. We are saying precisely that there is *no* such woman. When we say that Cinderella was poorly off until she met her Fairy Godmother, we are saying that *if there had been* such a woman—a

---

2. Stephen (1896).

woman fitting these descriptions—then this woman *would* have been poorly off until she met her Fairy Godmother.[3]

Suppose all this is granted. A defender of RKF may reply: If one's having an existence cannot be good for one, then one's having an existence cannot be *bad* for one either. If this is so, however, then RKF, while it may not be good for any of the animals that it raises and kills for food, cannot be bad for any of these animals.

However, it is crucial to distinguish between two parts of RKF:

1. RKF's bringing animals into existence, and
2. RKF's treating these animals in a particular way.

It is not (1) but (2) that is bad for animals. RKF is not bad for animals by bringing these animals into existence. RKF is bad for animals *by giving worse lives to these animals it has brought into existence than these same animals might have had*.

To this, a defender of RKF may object that I am assuming that RKF involves *factory farming*—that is, farming in which animals are raised in cramped spaces, caused to feel much pain during their lives, and killed in a brutal manner. While it is true that factory farming gives worse lives to animals than these same animals might have had, RKF need not involve factory farming. RKF might instead involve only *free-range farming*, which, let us say, gives to the animals in question happy lives—including, for example, plenty of green space to roam around in, good quality food, contact with each other, and so on—and then kills them painlessly in their sleep without their anticipation.

However, I am not assuming that RKF involves factory farming. Even if RKF were to involve only free-range farming, the lives it would give to the animals in question would still be much *shorter* than the lives they might

---

3. Some might worry that I have missed Stephen's point. His point, it may be said, is not that RKF is good for particular animals, but that it is good for particular *species* of animals. If there were no more pigs, then this would be bad for pigs, taken collectively. But it seems very hard to make sense of the idea of things going well or poorly for a species—independently, at least, of how things are going for particular members of this species. Moreover, even if there were a sense in which pigs as a species could do well independently of the well-being of particular pigs, it is unclear why their well-being in this sense would be normatively significant in the slightest. It is highly plausible that it is the well-being only of *individual* beings that has this sort of value. If no individual being is benefited by a given practice, then it is irrelevant that there may be some sense in which this practice is good for its species.

otherwise have. These animals would be much better off living longer lives in their free-range farms.

A defender of RKF may deny that animals such as cows, pigs, chickens, and other such animals have anything to gain from living longer. More life, it may be said, is good for a being only if this being *desires to live longer*, or at least has some *long-term plans, projects, or goals* that would be completed or fulfilled if it were to live on. Cows, pigs, chickens, and other such animals have no such desires, plans, projects, or goals, and so nothing to gain from additional life.

Some animal advocates have responded by claiming that cows, pigs, chickens, and other such animals *do* have such desires, plans, projects, or goals. As evidence of this, they have pointed to such things as the concern such animals seem to have for the survival and flourishing of their own offspring.

But this seems to me the wrong response, for two reasons. First, even on the most plausible desire-based theories of well-being, it is not the satisfaction or frustration of one's *actual* desires that is good or bad for one, but only those desires that one *would have if one were suitably idealized*—that is, a fully informed, vividly imagining, maximally mature version of oneself.[4] Cows, pigs, chickens, and the like might not, *as they are*, have any desires to live longer or for future things, but they might well have such desires if they were suitably idealized. It is common for families to speculate on what their family dog or cat might be like if he or she were to become much more intelligent and able to converse with them. Animals like dogs and cats seem to many of us to have individual personalities that might not only survive, but perhaps be made fully manifest by, their transformation into beings with greater cognitive faculties. If this were so, then we might expect such transformed beings to have preferences concerning how the lives of their actual, non-idealized selves are to go. Among other things, we might expect them to prefer longer rather than shorter lives for their actual selves (providing, of course, that these lives were to be lived on free-range farms).

Second, it does not seem necessary in order for additional life to be good for one that one have desires (actual or idealized) that would be satisfied by it. On the contrary, it seems enough that the additional life *would involve certain kinds of pleasures for one*. Why would more life be good for a normal human adult like myself? One reason is that there are certain kinds of pleasures on the horizon for me. Why would more life be good for a young or

---

4. See Sidgwick (1907) and Rawls (1971).

middle-aged cow, pig, or chicken living on a free-range farm? One reason, similarly, seems to be that, in such a setting, there are certain kinds of pleasures on the horizon for it.

I say *certain kinds of pleasures*, rather than simply *additional pleasures*, for an important reason. On a view I am tempted by, *purely repeated pleasures*—that is, pleasures that introduce nothing qualitatively new in terms of pleasurableness into a being's life—add nothing to that being's lifetime well-being.[5] Any further pleasures involved in a longer life are good for one only if these pleasures *bring something qualitatively new in terms of pleasurableness to one's life*. If all a person gets in having more life is just the same pleasures of watching their favorite sitcom over and over again, then this person has gained nothing by living longer. Similarly, I believe, if all an animal gets in having more life is just the same pleasures of chewing grass over and over again, then it has gained nothing by living longer. The thing is, however, I believe there is considerable scope for further qualitatively new pleasures in the life of a free-range animal who is still young or middle aged. My suspicion, indeed, is that many of those who believe that additional life cannot be good for an animal believe this only because they are falsely assuming that the only future pleasures available to cows, pigs, chickens, and the like—or at least cows, pigs, chickens that have spent some time in a free-range farm—are purely repeated ones.

What pleasures do I have in mind? Consider a family dog, Gertie. Imagine Gertie running around today in the local park, chasing sticks, meeting new dogs, having new olfactory pleasures, and exploring parts of the park she has never been to before. It seems clear to me, and I hope to you, that it was a good thing for Gertie that she lived on until today. If she had died peacefully in her sleep last night, this would have been bad for her, since she would not have lived on to experience all these wonderful qualitatively new doggy pleasures. Similarly, cows roaming free in a green paddock with plenty to eat, even if they have no future-oriented desires, may have evolving social lives with each other that are a source of qualitatively new pleasures for them as time goes by, or slow dawning realizations about their lives or vague increments in understanding that are pleasurable in various ways, or different or deeper appreciations of the field in which they are grazing as it undergoes changes during the shifting seasons, or experiences of watching their offspring grow into adulthood and reproduce themselves that involve pride or satisfaction.

---

5. For further defense of this idea, see Bramble (2015).

To kill them when they are young or middle aged would be to rob them of these qualitatively new pleasures.

I conclude that animals on free-range farms would be better off living into old age than being painlessly killed in their sleep during youth or middle age.

One final objection: What about RKF involving only free-range farming that allows the animals in question to live into their old age or until they die of natural causes?

I accept that RKF of this kind would not be bad for the animals in question. But would there be much of a market for such meat? This is unclear given that many consumers of meat seem to be of the opinion that the flesh of older animals is tough and tasteless. In any case, as I will be arguing in the next section, even RKF of this kind may be bad for *us*.

Of course, the crucial question is whether the amount that animals would gain by living into old age on free-range farms (rather than being killed in youth or middle age) is *greater* than the amount that humans would gain by these animals being killed in their youth or middle age (rather than when they are old). To answer this question, we need to know in what ways meat consumption affects human well-being. It is to this matter I now turn.

The Second Premise

Many people believe that human beings need to eat meat in order to be healthy. But even the American Dietetic Association acknowledges that

> vegetarian diets, including total vegetarian or vegan diets, are healthful, nutritionally adequate, and may provide health benefits in the prevention and treatment of certain diseases.[6]

Still, many meat-eaters remain unconvinced. They say they feel lacking in energy, and in health more generally, when they don't eat meat, and they take these feelings to be more reliable indicators of their levels of health than the findings of current science.

---

6. This is from the abstract of "Position of the American Dietetic Association: Vegetarian Diets," Journal of the American Dietetic Association, volume 109, issue 7 (July 2009), pp. 1266–1282, http://www.eatright.org/cps/rde/xchg/ada/hs.xsl/advocacy_933_ENU_HTML.htm.

But such feelings do not necessarily indicate ill health. They may simply be withdrawal symptoms from giving up a substance to which one has become addicted. Indeed, these feelings seem similar to those many of us have when we give up, say, coffee, and nobody thinks that the fact that one feels this way in the coffee case shows that giving up coffee is bad for one's health. (On the contrary, it is a feeling one must go through in order to regain health). Moreover, the meat industry has paid big money to advertisers to try to get us all to think of meat as necessary for health and vitality. We must factor this in when assessing our own feelings about whether meat is really necessary for our health.

It may be objected: But when I stop eating meat, I feel lethargic. This may not be evidence that my health is independently damaged, but this lethargy itself constitutes a decline in health. Health includes things such as how energetic or vital one feels.

However, even if this is true, withdrawal symptoms like these do not last very long. Persist with a vegetarian diet and one will likely soon feel energetic again—even more energetic than previously, depending on how much meat was in one's diet. One's cravings for meat, too, will disappear, or at least diminish substantially. All this would happen more rapidly still if meat were not readily available or if most others were also abstaining. Moreover, if our society were to give up meat, then future generations would not get addicted to meat in the first place and so suffer none of the withdrawal symptoms of having to give it up.

Suppose all of this is granted. Still, it may be objected: Meat consumption makes a very large contribution to our well-being because of the *pleasures* it gives us.

Many people seem to think that if we were limited only to vegetarian meals, eating would soon become a tiresome exercise, and life would lose much of its appeal.

However, as many vegetarians have pointed out, those who make this objection cannot have sampled very much vegetarian cuisine. While it is certainly true that the vegetarian options at most restaurants and fast-food joints today are pretty bland or unappetizing, they are hardly representative of what can be done in the kitchen without meat. Vegetarian meals can be not only healthful, they can be delicious and satisfying, and admit of such great variety that one need never get sick of them. Moreover, if we all stopped eating meat, the vegetarian options at restaurants would quickly get tastier and more varied, and in any case it is easy to learn how to make delicious vegetarian meals at home.

A more sophisticated argument has recently been offered by Loren Lomasky. Lomasky claims that the pleasures of meat

> afford human beings goods comparable qualitatively and quantitatively to those held forth by the arts. Lives of many people would be significantly impaired were they to forgo carnivorous consumption.[7]

Lomasky argues for this claim by appeal to the widespread and powerful human desire for meat (or its subjective importance to us). He observes that "All across the globe . . . as incomes increase so does the amount of meat in people's diets."[8] He goes on:

> When we look at the world's great cuisines we discover that almost without exception they not only include meat but also feature it as a focal point of fine meals. In France as in India, China as in Italy, meat is sovereign. . . . That so many religions advance constraints on which animals are to be eaten and how the permissible ones are to be slaughtered and prepared conveys a recognition of meat eating as being among the very important components of how human beings can live well.[9]

There are two problems, however, with this argument. First, it relies on a desire-based or subjective theory of well-being, and, as I suggested above, the most plausible versions of such theories hold that it is not one's *actual* desires, but only one's *idealized* desires, that determine what is good or bad for us. This is because, as David Sobel nicely puts it, idealized desires "are more fully for their object as it really is rather than for the object as it is falsely believed to be."[10] This is a problem because, while it may be true that most people have a very strong desire to eat or enjoy meat, it is not clear that they would continue to have this desire if they were suitably idealized. In fact, it seems likely that, apprised in a vivid way of all the gory details of the manner in which most animals who end up on our plates are raised and killed, most of us would not want to eat meat—let alone enjoy it—ever again.

---

7. Lomasky (2013), p. 190.

8. Lomasky (2013), p. 185.

9. Lomasky (2013), p. 185.

10. Sobel (2011), p. 59.

Second, as I also suggested above, desire-based theories of well-being are implausible. There is not space here to fully make the case against such theories, but recall my claim above that such theories cannot explain the value for us of future pleasures. It seems good for Gertie the dog that she lived on until today to experience an array of qualitatively new doggy pleasures, even if this involved no desire of hers being satisfied. If desire-based theories are false, then even if our desires to eat or enjoy meat were to survive idealization, this would not show our eating or enjoying it to be significantly good for us.

There is, however, a different, and better, way to argue for Lomasky's claim. This is to say that the pleasures of meat are extremely good for us, not because we want (or would want) them, but just because of the particular *phenomenology* of these pleasures themselves (i.e., "what it is like" to experience them). No vegetarian diet (at least given present technologies) is able to provide this particular pleasurable phenomenology. Any life without such phenomenology is to that extent impoverished.

Moreover, it may be added, if we all stopped eating meat, then our *cultures* would be greatly diminished, and along with these the richness of our social and cultural encounters. One way we stay connected to our ancestors is through the meals they pass down to us. If we all stopped eating meat, then this important link with the past would be severed.

What should we make of this argument? I think we have no choice but to accept that the pleasures of meat, and what these contribute to various social and cultural aspects of our lives, make us well off in ways that no quantity or quality of vegetarian food could possibly achieve (again, given present technologies). The absence of these pleasures from a person's life (even in the life of someone who has no desire to eat or enjoy meat) represents a real loss for that person. Moreover, this is not a trivial loss. These pleasures are *significantly* good for one.

However, the important question is: Is this significant loss a significant *net* loss? In what follows, I will sketch three reasons for thinking that it is not.

First, what we would lose by giving up the cultural traditions associated with our meat consumption may be fully compensated for by pleasures gained from reinventing these dishes in vegetarian ways and forging new traditions at the dinner table. It is not as if by removing meat from our diet *all* our important connections with the past would be severed. There would still be a great deal of cultural continuity that would be possible. And we should not underestimate what may be gained by starting afresh and exercising our creativity.

Second, meat is very costly to produce. If we were to cut meat from our diets, the resources that are currently spent on its production could be redirected toward other areas of our lives such as health, education, infrastructure, and so on.

Third, there are reasons to believe that there could be some heavy psychological costs associated with meat consumption. Consider, first, that most of us grow up as children who love animals. When we first discover that the meat on our plate is the body of an animal who has been killed for our consumption, this distresses us greatly. When we learn further of what goes on in farms and slaughterhouses, even free-range ones, many of us are truly horrified. We are, however, very good at putting these thoughts out of our heads and carrying on with our meat-eating—especially given the often considerable social and economic pressure to do so. But an idea ignored can continue to affect one. There is a growing body of evidence that most of us experience many kinds of significant pleasurable and unpleasurable feelings *without being aware of them*—that is, in the background of our consciousness.[11] An especially vivid example (on the pleasure side) is provided by a patient of Oliver Sacks, who writes:

> Sense of smell? I never gave it a thought. You don't normally give it a thought. But when I lost it—it was like being struck blind. Life lost a good deal of its savor—one doesn't realize how much "savor" is smell. You smell people, you smell books, you smell the city, you smell the spring—maybe not consciously, but as a rich unconscious background to everything else. My whole world was suddenly radically poorer.[12]

On the side of unpleasurable experiences, Daniel Haybron writes:

> Some affective states are more elusive than the paradigmatic ones, particularly moods and mood-like states such as anxiety, tension, ennui, malaise. . . . They may exceed our powers of discernment even while they are occurring. . . . A vague sense of malaise might easily go unnoticed, yet it can sour one's experience far more than the sharper and more pronounced ache that persists after having stubbed one's toe. Likewise for depression, anxiety and related mood states, at least in their milder forms. Consider how a tense person will often learn of it only when receiving a massage, whereas stressed or anxious individuals

11. See, for example, Haybron (2007), Schwitzgebel (2008), and Bramble (2013).

12. Quoted in Rachels (2004), p. 225.

may discover their emotional state only by attending to the physical symptoms of their distress. Presumably being tense, anxious, or stressed detracts substantially from the quality of one's experience, even when one is unaware of these states.[13]

How can this happen? Haybron explains it as follows:

> Everyone knows that we often adapt to things over time: what was once pleasing now leaves no impression or seems tiresome, and what used to be highly irritating is now just another feature of the landscape. Could it also be that some things are lastingly pleasant or unpleasant, while our awareness of them fades? I would suggest that it can. Perhaps you have lived with a refrigerator that often whined due to a bad bearing. If so, you might have found that, with time, you entirely ceased to notice the racket. But occasionally, when the compressor stopped, you did notice the sudden, glorious silence. You might also have noted, first, a painful headache, and second, that you'd had no idea how obnoxious the noise was—or that it was occurring at all—until it ceased. But obnoxious it was, and all the while it had been, unbeknownst to you, fouling your experience as you went about your business. In short, you'd been having an unpleasant experience without knowing it. Moreover, you might well have remained unaware of the noise even when reflecting on whether you were enjoying yourself: the problem here is ignorance—call it reflective blindness—and not, as some have suggested, the familiar sort of inattentiveness we find when only peripherally aware of something. In such cases we can bring our attention to the experience easily and at will. Here the failure of attention is much deeper: we are so lacking in awareness that we can't attend to the experience, at least not without prompting (as occurs when the noise suddenly changes).[14]

Similarly, I want to suggest, it may be the case that, knowing what meat is—and, in particular, what we do to animals in farming and slaughtering them for food—sours or pollutes our experiences of eating meat, and perhaps our experiences of living in this world more generally, in ways that are very hard or

13. Haybron (2008), p. 202.

14. Haybron (2007), p. 400.

even impossible to attend to while we are still meat-eaters. Certainly, many people claim to find the experience of giving up meat similar in various respects to the experience Haybron describes of being at home in one's kitchen when the compressor of the whining refrigerator switches off. Many report that they experienced a tremendous sense of relief or freedom, or a lightness of being, after giving up meat—feelings they had not anticipated, and that suggest they were experiencing unconscious pain beforehand.

Part of the unconscious pain felt by meat-eaters, I suspect, has to do with their having deliberately turned away when they were children from something that they sensed at the time was an important moral issue. This turning away seems likely to leave one with a burden comparable to that carried by a person who has reason to suspect a friend of theirs of having committed some heinous crime, but who refuses to investigate further or turn her friend in for some relatively trivial reason (say, fear of upsetting the balance of her social life). People who ignore qualms they have or silence parts of themselves cannot be entirely happy individuals. Moreover, their being like this may prevent them from being the sort of open people who are able to take joy in many other aspects of life. So, ignoring the issue of meat, refusing to investigate, may close one off to various other possible pleasures.

It may be objected: But what about *free-range* farming? If (as I conceded in the first section) animals who are raised in free-range farms and get to live on into old age are not harmed by RKF, then why should we have any qualms at all about participating in this system that raises and kills them for food? Why should our participation in such a system have any tendency to make us feel bad?

I accept, of course, that there would be nothing bad about such a system deriving from harms inflicted on these animals. In such a system, these animals are not harmed, and so there can be nothing of disvalue deriving from their being harmed. Since this is the case, there may be a sense in which we *should* not be disturbed or upset by such a system, or by our participation in it. But even such a system, I want now to suggest, would nonetheless cause most of us psychological suffering (even if we believed it should not).

Let me explain. For most of us, the thought of the dead bodies of our friends and loved ones, or even those of complete strangers, being cut up or torn to pieces is deeply distressing. It is even worse to think of their body parts then being devoured by some creature. That it causes such distress to us seems to be, not because we think it harms these people (for they are already dead and so cannot be harmed by anything anymore), but because it reminds us of the fact that we are embodied (and so finite) beings, and with that much of the suffering and tragedy of our lives. We prefer to bury intact the bodies of our loved ones, or burn them, so that it is not possible for them to be taken apart.

I suspect that, for many of us, there is a similar pain—albeit often an unconscious one—that accompanies our thoughts of what takes place in slaughterhouses, even slaughterhouses where the animals in question have been killed painlessly without their anticipation. These practices are unavoidably grisly. The thought of the bodies of these animals being taken apart, ending up on our plates, and being devoured by us, is a painful reminder of what we all are: embodied beings prone to disease, suffering, and death.[15]

To emphasize: I do not pretend to have proven here that we suffer any of the unconscious pains I have been describing. What I have said remains largely speculative. But I do hope to have persuaded you that there is some possibility, and perhaps also some reason to believe, that such pains exist— that is enough for my purposes. Whether such pains actually exist I will leave to the scientists of the future, with their superior technologies, to confirm or disconfirm.

I conclude that, while the absence of the pleasures of meat in a person's life truly does represent a significant loss for that person, when we take into account (i) the pleasures involved in forging new culinary traditions, (ii) the economic opportunity costs of meat production, and (iii) the possible psychological costs associated with meat-eating, we see that it is probably not a *significant net loss*, and may not even be a net loss at all.

The Unimportance of Animal Well-Being

According to some, even if RKF is extremely bad for animals and only trivially good for us, RKF should continue. This is because animal well-being is far less important than human well-being. Great harm to animals is less bad than the relatively small sacrifice involved for us in giving up meat.

But this is an implausible idea. There seems no good reason why the well-being of some creatures should be worth more than the well-being of others. It seems far more plausible to think that the intrinsic value *simpliciter* of some increase in a being's lifetime well-being is proportional just to *the amount of the increase*.

---

15. For a summary of some recent studies examining the relationship of a vegetarian diet to emotional well-being, see Ruby (2012), p. 146.

Why, then, do some people *think* that human well-being is worth more than animal well-being? As others have pointed out, it seems likely to have to do with the fact that humans seem capable of having much higher levels of lifetime well-being than any other animals on this planet. Those who think human well-being is more valuable may be confusing the fact that we can be more greatly benefited with our benefits having greater value.

For this diagnosis to be right, however, we need an account of why human beings are capable of much higher levels of lifetime well-being than animals. No one has yet provided a satisfactory such account. In the remainder of this section, I want briefly to suggest one.

The reason human beings are capable of much higher levels of lifetime well-being, I believe, has to do with a point I made in the first section, namely that purely repeated pleasures add nothing to a being's level of lifetime well-being. Human beings have available to them *much greater diversity* in pleasurable experiences than other animals do. While I claimed above that Gertie the family dog may experience many qualitatively new pleasures on a given day in exploring the park, meeting other new dogs, smelling new smells, and so on, I think these pleasures are quite limited when compared with the pleasures we humans are able to obtain from our much deeper relationships with each other, much greater capacity to understand ourselves and learn about the world, more sophisticated experiences of art and beauty; ability to appreciate the importance of things, set goals, and work toward their completion; and capacity for selfless or virtuous behavior.

It is an easy mistake to confuse this greater capacity for well-being with our well-being's having greater value. But it is a mistake. The death of a normal cow in a paddock is less bad than the premature death of a normal human being, but this is not because human well-being matters more than animal well-being. It is because there is much more that continued life can add to the lifetime well-being of a normal human being than to the life of a cow. When we *do* genuinely harm animals a lot—as RKF does—this *is* extremely bad.

Causal Impotence

Suppose everything I have said so far is correct, and it would be best if RKF were to stop. Nonetheless, it may be claimed, none of us has a reason to abstain from eating meat. This is because our individual purchasing decisions have only a negligible effect on the demand for meat, and so none at all on RKF.

Strictly speaking, this objection does not threaten the conclusion of The Case Against Meat. After all, this conclusion says nothing about our individual reasons to act. It says only that RKF should be stopped (or, as I have been taking this to mean, that it would be *best* if RKF were to come to an end).

Nonetheless, I believe that even if we cannot, by abstaining from meat, improve the lives of any animals, we may each have sufficient reason to abstain from it. This is because, as I suggested in the second section, it is possible that there are heavy psychological costs associated with meat-eating. While the absence of the pleasures of meat in a person's life represents a significant loss for that person, it is a loss that may be fully compensated for by freedom from these psychological costs. If this is right, then many of us may have most *self-interested* reason to stop eating meat.

Furthermore, even if most of us lack the power to cause many other people to give up meat (and so reduce demand for meat enough to save or improve any animal's life), most of us are able, through making changes to our own diet and publicly opposing RKF, to cause some of our friends to abstain from meat as well, which would be very good for *them*.

Conclusion

In this chapter, I have tried to shore up The Case Against Meat by offering new responses to the four main objections to this argument. In the first section, I argued that RKF is bad for animals, not by bringing them into existence, but by giving worse lives to the animals that it has brought into existence than these same animals might have had. I argued that even free-range farming is bad for animals by giving them shorter lives than they would have had if they had lived on into old age.

In the second section, I argued that while the absence of the pleasures of meat in a person's life represents a significant loss for that person, it is not a significant net loss, and may not even be a net loss at all in light of the opportunity costs of meat-eating and the possibly heavy psychological costs associated with meat-eating.

In the third section, I tried to explain why human beings are capable of higher levels of lifetime well-being than other animals on this planet, and so why some people might mistakenly believe that human well-being is more valuable than animal well-being.

Finally, in the last section, I pointed out that it is no objection to The Case against Meat if, as individuals, we cannot save or improve any animal's life by becoming vegetarian. I then suggested that, even if we cannot save or improve any animal's life by becoming vegetarian, we may have sufficient self-interested reason to stop eating meat due to the psychological costs of meat-eating I described in the second section.

We should stop raising and killing animals for food. Our practice of doing so is very harmful for these animals and is not very good—and perhaps even, on balance, bad—for us.

References

Bramble, B. 2013. The Distinctive Feeling Theory of Pleasure. *Philosophical Studies*, 162, 201–217.

Bramble, B. 2015. On Susan Wolf's "Good-for-Nothings". *Ethical Theory and Moral Practice* (online first).

Haybron, D. M. 2007. Do We Know How Happy We Are? On Some Limits of Affective Introspection and Recall. *Nous*, 41(3), 394–428.

Haybron, D. M. 2008. *The Pursuit of Unhappiness: The Elusive Psychology of Well-Being*. Oxford: Oxford University Press.

Lomasky, L. 2013. Is It Wrong to Eat Animals? *Social Philosophy and Policy*, 30 (1–2), 177–200.

Rachels, S. 2004. Six Theses about Pleasure. *Philosophical Perspectives*, 18(1), 247–267.

Rawls, J. 1971. *A Theory of Justice*. Cambridge, MA: Harvard University Press.

Ruby, M. B. 2012. Vegetarianism. A Blossoming Field of Study. *Appetite*, 58, 141–150.

Schwitzgebel, E. 2008. The Unreliability of Naive Introspection. *Philosophical Review*, 117, 245–273.

Sidgwick, H. 1907. *The Methods of Ethics*. London: MacMillan and Company.

Singer, P. 1975. *Animal Liberation*. New York: HarperCollins.

Sobel, D. 2011. Parfit's Case against Subjectivism. *Oxford Studies in Metaethics*, 6.

Stephen, L. 1896. *Social Rights and Duties*.

III FUTURE DIRECTIONS

9 VEGANISM AS AN ASPIRATION

Lori Gruen
Robert C. Jones

How you cling to your purity, young man! [. . .] All right, stay pure! What good
will it do? [. . .] Purity is an idea for a yogi or a monk [. . .] Well, I have dirty
hands. Right up to the elbows. I've plunged them in filth and blood.

—HOEDERER FROM SARTE'S *DIRTY HANDS*

Introduction

Most people are now aware of the extreme suffering routinely experienced by animals raised for consumption in industrialized meat
and dairy production facilities.[1] Undercover video of some of the
most egregious forms of cruelty have made their way to the public.
In response, the industry has begun promoting "ag-gag" legislation,
anti-whistleblower laws that criminalize photographing or video
recording inside these facilities. Despite desperate efforts to conceal how animals are treated, the cruel everyday practices found on
factory farms and in slaughterhouses are no longer the industry's
dirty little secrets. Although there is increasing awareness of the
horrible conditions that animals endure, the vast numbers and the
extent to which these practices impact our shared world remain relatively obscure.

---

1. Much of this discussion draws on Gruen's previous publications. See Gruen 2014 and 2011.
We would like to thank Gunnar Theodor Eggertsson for important conversations that contrib-
uted to thinking about these issues. We would also like to acknowledge the generally instruc-
tive conversations that occurred at the Animals and Society and Wesleyan Animal Studies
(ASI-WAS) Summer Fellowship Program in 2012, which helped us shape this chapter.

According to the USDA Foreign Agricultural Service, roughly 1.02 billion cattle, 1.2 billion pigs, and 40 billion chickens worldwide are raised for food, most on factory farms (or what is referred to in the industry as Concentrated Animal Feeding Operations—CAFOs). Most of these animals are confined indoors for their entire lives in areas that prevent them from moving around; they are denied species-typical social interactions, including raising young, who are removed at birth; and they are subjected to a variety of painful procedures—tails and ears are cut off and males are castrated without anesthesia, animals are branded with hot irons, birds have their beaks sliced off with hot knives, in the egg industry male chicks are ground up alive, and dairy cows are forcibly impregnated regularly to produce milk. Though the normal lifespan of a chicken is approximately 10 years, laying hens are "spent" and unable to produce eggs after just 2 years, at which time they are slaughtered. Broiler chickens are genetically modified so as to grow to "processing" weight in only 6 weeks, at which time they are sent to slaughter. Slaughter often doesn't bring immediate relief from suffering as animals are shackled at the feet, hung upside down on a conveyer belt, and only occasionally are their throats slit cleanly enough that their deaths are instantaneous, leaving many to linger in pain, bleeding until they lose consciousness.

These numbers don't include sentient beings who live in the sea. One source puts the number of marine animals killed for food in the United States alone at 51 billion (FFH, 2011). Common aquaculture procedures include taking animals out of their water environments, asphyxiating them in ice or in CO_2-saturated water, and cutting their gills.

As if the magnitude of animal suffering wasn't enough to cause reasonable people to pause and consider the pain and death they contribute to in order to satisfy their personal tastes, industrialized food production is responsible for unprecedented damage to the environment, damage that harms humans and other animals. In the United States alone, the cattle, pork, and poultry industries produce nearly 1.4 billion tons of animal waste, 130 times the amount of waste produced by the entire human population of the United States (USSCANF, 1997). These wastes end up in our waterways and underground aquifers. In addition, the antibiotics fed to livestock produce antibiotic-resistant bacteria and resistance genes that make their way into ground and surface water, causing public health concerns. Most alarmingly, the UN conservatively estimates that roughly 18% of the total greenhouse gases emitted come from industrialized livestock production, more greenhouse gas emissions than all the transport on earth—planes, trains, and cars—combined (Steinfeld, et al., 2006).

The costs in terms of the violence, suffering, exploitation, domination, objectification, and commodification of animals for food, the destruction of the environment and the displacement of animals in the process, as well as costs to our own health and the health of the planet, call for immediate, effective, and decisive action at the personal, collective, and policy levels. We support ethical veganism as an empowering response to these atrocities. Ethical veganism is a commitment to try to abstain from consuming products derived from animals including meat, dairy, and eggs, as well as products derived from or containing animal products as an ingredient. In this chapter we discuss two different ways that people conceive of veganism,[2] veganism as a lifestyle or identity, and veganism as a goal. We argue that there are conceptual and practical problems with the former, explore arguments about whether either actually makes a difference, and optimistically conclude that veganism as an aspiration can.

Two Senses of "Vegan"

Many ethical vegans sincerely adopt veganism as a *lifestyle* as an expression of their commitment to ending the suffering that accompanies the commodification of sentient beings. Ethical vegans often see themselves in solidarity with one another in the struggle against cruelty and violence. Often the idea of veganism is accompanied by a sense that those practicing it have achieved a kind of ethical purity. Once one adopts a vegan lifestyle, she then has "clean hands" and may carry on her consumerism with a clear conscience, since no animals were harmed in the production of her vegan consumer goods. Sometimes seen as a kind of litmus test of one's commitment to social justice for animals, veganism is often thought to be the "moral baseline" for those seeking to end the suffering and domination of other animals.

Though there are debates among vegans about questions of purity and commitment, there appears to be a growing public perception of vegans—that may be based in fact, prejudice, or more likely a combination of both—that vegans see themselves as better than and morally superior to non-vegans; that they can be "preachy," and even annoying; that they often exhibit a kind of self-righteous zealotry, acting as the "vegan police" who promulgate veganism as the universal, one-and-only way to fight systemic violence against animals. Often these vegans are thought to judge non-vegans, including ovo-lacto vegetarians, as shirking their responsibility or being self-indulgent or simply cruel.

2. Other forms of veganism are discussed in the literature as well, for example, "veganarchism," "boycott veganism," and "engaged veganism." See for example Dominick 1997 and Jenkins & Stănescu 2014.

This view, that the only ethical way to live is to adopt a vegan lifestyle, we call *Identity Veganism* (V_I). If followed strictly and universally, V_I is thought to keep one's hands clean. As the name implies, this sort of veganism is often thought of as an *identity*, and some people who would fall under V_I have even claimed that they are discriminated against as vegans. These vegans have an air of moral certitude and moral superiority. It was perhaps proponents of V_I that prompted philosopher Val Plumwood to describe vegans as "crusading [and] . . . aggressively ethnocentric, dismissing alternative and indigenous food practices and wisdom and demanding universal adherence to a western urban model of vegan practice in which human predation figures basically as a new version of original sin, going on to supplement this by a culturally familiar methodology of dispensing excuses and exemptions for those too frail to reach their exacting moral norms of carnivorous self" (2000, p. 286).

Of course, the V_I lifestyle we are describing comes in degrees. But there is another sort of veganism, what we will call Aspirational Veganism (V_A), that views veganism not as a lifestyle or identity, but rather as a type of practice, a process of doing the best one can to minimize violence, domination, and exploitation. On this view, veganism is an *aspiration*. V_A commits us to striving for a moral goal; V_A is something that one works at rather than something one is. Rather than seeing veganism as a kind of universal norm to be imposed as a moral imperative, on this view we should instead see veganism, as ecofeminist philosopher Marti Kheel suggests, as an invitation in response to the violence, exploitation, domination, objectification, and commodification that sentient beings endure in modern industrialized food production processes, part of a larger resistance to such harm and destruction (Kheel, 2004).

We don't think that every vegan is always either a V_I or V_A; there is certainly some overlap here, and in different contexts someone who recognizes veganism as an aspiration may also express her commitments in ways that make it seem more like a lifestyle. Importantly, both types of vegans oppose the systematic cruelty toward and destruction of other animals. However, to see veganism as an aspiration is not to see veganism as *merely* an aspiration. To call oneself a vegan in the V_A sense while continuing consciously to act in ways that condone animal exploitation (for example, continuing to eat meat) would be to disingenuously appropriate the language of V_A and act in "bad faith." The focus of V_A is to imagine and earnestly try and actualize—to the best of one's ability—a world in which there is no animal exploitation, by working to minimize violence. While this is also a goal that V_I shares, to ascribe moral purity and clean hands to veganism is to make a category mistake. In the next section we discuss why we believe this to be the case.

Why Veganism Can Only Be an Aspiration

The belief that a rejection of industrialized livestock products allows one to avoid complicity in harming other animals is too simplistic and ignores the complex dynamics involved in the production of consumer goods of all kinds, global entanglements we engage with each time we purchase and consume food of all sorts. Vegan diets have "welfare footprints" in the form of widespread indirect harms to animals, harms often overlooked or obscured by advocates of V_1. Industrialized agriculture harms and kills a large number of sentient field animals in the production of fruits, vegetables, and grains produced for human (not livestock) consumption. As MacClellan notes, "large farm equipment used in the industrial agricultural production of staple crops such as wheat, corn, and soybeans harms many sentient field animals, including members of many species of rodents such as mice and voles, as well as rabbits and birds," not to mention reptiles and amphibians (forthcoming, p. 12).

Despite wanting it to be otherwise, vegan or not, we cannot live and avoid killing. Living today, even for vegans, involves participating unwittingly in the death of sentient individuals. For example, animal products are found in or used in the production of a great number of consumer goods including auto upholstery, beer, bread, candles, chewing gum, cosmetics, cranberry juice, deodorants, fertilizers, hairspray, house paint, lipstick, marshmallows, nail polish, plywood, perfume, photographic film, pickles, pillows, red lollipops, rubber, sauerkraut, shaving brushes, shaving cream, soap, soy cheese, sugar, surgical sutures, tennis rackets, transmission fluid, vitamin supplements, and wine.[3] We can rail against the massive violence that is done to the huge number of living beings who did nothing to deserve their tragic fates, but neither our political commitments nor our moral outrage place us above the violent fray. All aspects of consumption in late capitalism involve harming others, human and nonhuman.

One of the most troubling examples is palm oil, a ubiquitous ingredient found in a large number of prepared "vegan" food products. Produced by clear-cutting, palm oil plantations in Southeast Asian countries such as Borneo and Sumatra have nearly wiped out remaining orangutan populations while harming members of many other endangered (and nonendangered) species. As demand for palm oil grows and as new plantations

---

3. There are other products too. This is part of a list published by PETA, http://www.peta.org/living/beauty/animal-ingredients-list/.

are developed in Africa, the destructive impact of palm oil may be greater than that of some products made directly from animal bodies or bodily excretions (Hawthorne, 2013).

Vegans have attended to the tragedy that farmed animals experience, but have generally paid less attention to the harms other animals suffer in the production of vegan foods. Thinking about consumption in a time of climate change may provide a clearer way to understand the ripples of responsibility. Though it is hard to calculate the direct harms to humans and other animals from greenhouse gas emissions attributable to the agricultural sector, it is impossible not to contribute to these harms and still eat (Gruen and Loo, 2014). To be sure, vegan diets are less harmful than those that include animal products, but vegan diets are by no means "emissions neutral," and this is just one dimension upon which humans, vegan and non-vegan, negatively impact the earth and other animals. If we picture our responsibilities as a web, with direct harms at the center, vegans are certainly closer to the periphery than those who consume animal bodies, who kill animals, and who directly profit from the death of other animals, but vegans are still a part of the web, and not, as many practitioners of V_I seem to believe, beyond reproach.

Living necessitates dying and, controversially, killing. We can't live without killing others or, at best, letting them die. When we live with companion animals, for example, other animals will have to die, most obviously to feed those animals. Even if they are vegan, dogs and cats will kill and eat other animals if they get a chance. And when we deny them that opportunity, it becomes more obvious how problematic our power over them is. If we are all vegan, growing plants to feed ourselves and other animals involves killing some other animals. Even if some vegans can practice "veganic" farming, that is, carefully growing plants in such a way as to not harm or displace the animals who live on the land while growing enough food to share with the "denizens" that may raid the fields—the vast majority of us cannot afford to create food in this way (Gruen, 2014).

Given this, veganism can be but an aspiration, and imagining oneself to be V_I is an illusion.

"Humane" Killing

One might wonder whether aspiring to V_A condones purchasing locally raised animals who are "humanely" killed. Given that we are always implicated in the deaths of other animals, perhaps this recognition is what motivates young, affluent (mostly white) "students" with sizable disposable incomes to spend

$15,000 to enroll in courses like the twelve week "full-immersion" butchery program at Fleisher's Grass-Fed and Organic Meats in New York. Upon graduation, students are guaranteed to be able to butcher a lamb, pig, and steer. Tuition includes knives, "butcher's armor," and a copy of Fleisher's *The Butcher's Guide To Well-Raised Meat*. Students are also encouraged to read Michael Pollan, Joel Salatin, *The River Cottage Cookbook*, and *The Niman Ranch Cookbook*. Such do-it-yourself (DIY) "craft" butchery classes can be found in many other "foodie-friendly" cities such as Philadelphia and San Francisco. Underlying this booming alternative food movement is an increased awareness of the destructive nature of industrialized meat production, coupled with a sentimental nostalgia for a time when a majority of Americans were farmers and crafts persons living closer to the rhythms of the natural world. Described as locavorism, compassionate carnivorism, the sustainable meat movement, the humane meat movement, the happy meat movement, the nose-to-tail food movement, and the conscientious omnivore movement, these alternative food movements market themselves as "free range," "grass-fed," "organic," "natural," or "cage-free," all of which are thought to stand in for "humane."

Since it is difficult to deny the cruelty involved in industrial animal production, it is promising to learn that there are growing numbers of people who are wary of participating directly in agribusiness. The possibility of "happy meat" may seem to offer an ethical alternative to the cruelty of the factory farm, ensuring happier lives and "humane deaths" for animals destined to become meat. An interview with Joshua Applestone, owner of Fleisher's Grass-Fed and Organic Meats, exemplifies this core tenet of the "humane" meat movement:

Q: You were [a vegetarian]. What caused you to become [an omnivore]?

JOSH: After about 6 months of running Fleisher's it was our bacon that put me back on a meat-eating track. My vegan/vegetarianism was an outgrowth of my beliefs about how horrible the factory-farmed meat industry is. Once I really knew where my meat was coming from and how these animals were treated and slaughtered I could feel comfortable eating meat again (Applestone, 2011).

Measured against the vast majority of consumers who are completely disconnected from the suffering they cause when they buy neatly shrink-wrapped cuts of meat, compassionate carnivores deserve some praise. Yet despite their supposed concern for the well-being of animals, relatively little attention has been paid to the actual treatment of animals on "local" farms.

Sadly, animals in smaller operations sometimes suffer more acutely than animals raised in factory farms due to lack of consistent veterinary care, given that such care is expensive and time-consuming for small farmers. Bohanec (2014) argues that when it comes to "humane" versus factory-farmed meat, the similarities outweigh the differences. For example, so-called "cage-free" eggs come from hens who, like hens raised on factory farms, experience over-crowding, debeaking, and a terrifying slaughter. So-called "organic" dairy products come from cows who are artificially inseminated and kept pregnant their entire lives. Their calves are removed at birth, where male calves are sent to auction for use as veal or beef. "Humane" meat comes from animals that, as on factory farms, experience tail docking, ear notching, castration, tooth-filing, and de-horning, all without anesthesia.

And slaughter is often done in the same way it is for animals reared more intensively. An overwhelming majority of animals raised on local farms are sent to industrial slaughterhouses, killed alongside their kin raised in indus-trial operations. A small minority of pasture-based farmers take pains to ensure that the animals they raise are killed with respect. Tim Young, for ex-ample, found a processor an hour from his Nature's Harmony Farm that kills 9 cows a day, compared to the 400 an hour killed in large processing plants. Slowing down the killing process minimizes fear and helps to ensure that pain is minimized. When possible, Tim is present as the cows are killed. As he puts it, he wants to "be there to look each one of my animals in the eyes so that they can at least have a familiar face." It is also his way of paying his last re-spects (Gruen, 2011).[4]

In order to avoid forcing animals to endure the terror of transport to slaughter, another small group of farmers is hiring "mobile slaughterhouses" that come to the farm to kill and process the animals. These Mobile Slaugh-ter Units (MSUs) are USDA-approved slaughterhouses-on-wheels that travel to small farms, slaughtering animals on-site. One of these mobile units, owned by Lopez Community Land Trust in Washington, is a specially equipped, refrigerated trailer that is pulled to the farm by a diesel truck. After killing the animals (5–9 cows per day), the unit then drives the car-casses to a facility where they are cut into portions.[5] Elizabeth Poett, who operates an organic ranch in Rancho San Julian, California, is proud to use

4. http://www.naturesharmonyfarm.com/grass-fed-meat-farm-blog/2008/2/21/local-meat-processor.html.

5. Etter 2008.

an MSU. According to Poett, the MSU provides each of her 600 cattle with "more noble deaths and cut[s] out the need for a long final slog in the back of a trailer to a far-off killing floor. It's a dream to be able to run this beef business like I've been able to do it with the mobile harvest unit. I sleep better at night" (Adelman, 2009).[6]

Though MSUs slaughter fewer animals, they share more in common with industrial "processing" facilities than one might imagine. Animals are stunned with a captive bolt gun (or a firearm), sometimes taking two or three shots to render the animal unconscious. The animal's throat is then slit and the body hung to bleed out, be disemboweled, and dismembered (Bohanec, 2013).

Those who aren't ready to forgo consuming animals but who are uncomfortable with industrialized animal production often romanticize the connection they imagine they make to the dead animals they consume. Being involved in every step of production, including slaughter, creates a type of deep involvement that they can promote as laudable in an age when so much consumption is the result of various kinds of alienation. For many "compassionate carnivores," killing and eating animals is justified by their sense of respect for the connection they develop with the food they eat, where personally involving oneself in the death of an animal seems to provide a more direct and ethical way of eating, one that honors the subjects of slaughter while they are being consumed. Killing the animals one raises is thought to generate a sense of humility and remind people of our interdependence with other animals.

But this connection may be more rhetorical than genuine. One backyard chicken farmer who wanted to kill her rooster named Arlene describes her experience killing and preparing Arlene's body for consumption as being "as messy and mundane as cleaning the gutters."[7]

Indeed, slaughter often requires creating distance, not connectedness. Original Country Girl, a DIY butcher who, in giving advice to fellow DIY butchers on her blog, writes:

> The best advice is to always maintain a distance between you, and those intended for your dinner plate. This makes the butchering much

---

6. Regardless of size, meat producers commonly employ the term "harvest" to refer to the slaughter, disembowelment, and dismemberment (also known collectively as "preparation") of "livestock" and "poultry" (chickens are not considered livestock).

7. As cited in Gruen 2011 http://www.doublex.com/section/life/what-i-learned-when-i-killed-chicken.

easier if the animal is nothing more than "the black chicken" or "the grey and white goose." You can care for your critters in a humane and respectful way without allowing attachments to form. Rule number one is to never give *it* a name. Some people can get by with ironic names like the . . . "Christmas Dinner . . . " but for others even this can cause trouble later on. If you know you're soft-hearted don't do it. Clean the pen, feed good feed, and tend any wounds but don't get too close. No names, no handfed treats, and no special treatment for any one individual animal (McWilliams, 2011, emphasis added).

Katie Gillespie characterizes this as "connected disconnection." She writes, "[a]ll of the justifications for DIY slaughter . . . are enlisted to conceal what the process really does. DIY slaughter connects participants to the violence against the animal, and not to the animal him/herself. This 'connection' is a wholly false connection" (2011, p. 120).

Gruen (2011) argues that the problem with "humane farming" as well as industrial farming is that it relies on putting animals in the category of the edible, stripping them of their individual personalities and interests and viewing them as food. Being cruel to animals by causing them to suffer in factory farms is certainly objectionable. But animals have interests beyond suffering that matter as well—being allowed to live their lives with their family members and not being killed simply to satisfy someone else's culinary desires are some of those other interests. Even if other animals are raised "humanely," these interests are violated when they are slaughtered.

Edible Entanglements

Imagine how human interactions might be different if we saw each other as edible. If we allowed for the humane rearing of some humans for occasional consumption, this could lead to a breakdown in respect for one another and for humanity as a whole. We can already get a glimpse of the level of violence and disrespect that befalls those who are categorized as "disposable."[8] Being categorized as edible, in industrial societies, renders beings as consumable commodities. When we allow certain "things" to be bought and sold on the

---

8. Police shooting of black men along with mass incarceration rates in the United States, and the slaughter of genetically ill-suited zoo animals at European zoos are prominent examples of the dangers of categorizing others as "disposable."

market, we change the relationships we have and how we think of those relationships. We humans understand ourselves as not in the category of the edible, and this understanding, in part, shapes how we construct our relations with each other and the ways of life we share. If we now think of our bodies and other people's bodies as food, the value of our bodies and ourselves changes.

In response, it might be argued that since both human and nonhuman animals are, as a matter of fact, consumable, the problem is not that we ontologize animals as food, but that we ontologize animals as *meat*. Plumwood argues that refusing to allow sentient beings—including humans—to be categorized as edible leads to a rejection of ecological embodiment, since all embodied beings are food for some creature or another. Plumwood advocates a distinction between food and meat, where "meat" represents reductionism, domination, alienation, and commodification, while "food" suggests an acknowledgment of our ecological selves. As Plumwood puts it, "no being should be treated reductionistically as meat, but we are all edible (food), and humans are food as much as other animals, contrary to deeply entrenched beliefs and concepts of human identity in the west" (Plumwood, 2000, p. 295). To blind ourselves to this truth further distances and disconnects us from our ecological entanglements.

But Plumwood conflates the fact that we are all consumable with the fact that we categorize some bodies as "edible" and others as "non-edible." The fact that Plumwood almost became a crocodile's supper[9] and that all of us could be consumed as "prey" in certain contexts is an important recognition of our vulnerability. But this recognition is distinct from the social categorization of certain others as edible. To aspire to be vegan is not to deny ecological entanglement, but to suggest a reconceptualization of animals in their living bodies as fellow creatures with whom we can be in empathetic relationship and for whom we must have deeper respect (Gruen, 2015). V_A can provide a connection to other animals and the workings of nature by encouraging us to recognize the ways that our choices have far-reaching impacts.

While V_A might avoid the charge that it disconnects humans from the workings of nature, it is often argued that V_A is just one way, among many,

---

9. To read the horrifying details of her near-fatal encounter with a crocodile, see Plumwood (1996).

to honor our environmental entanglements. Even vegans, so the argument goes, cannot escape the cycle of industrialized violence and destruction of animals and their habitats. For example, one can exclude "animal products" from one's diet while including foods like tofu—made from soybeans, produced by Monsanto, using unsustainable, environmentally destructive monoculture practices—and still call herself a "vegan." Therefore, though caring, compassionate people have good reason to engage ethically with animals, there is no compelling reason to privilege veganism over other ways of being an ethical consumer. Protesting GMOs, spreading the word about the devastating impacts of palm oil production, or working to help forest animals whose habitats are being destroyed for raw materials used in the manufacture of cell phones are all just as important as going vegan.[10]

But one needn't choose to either try to forgo the products of direct violence on the one hand or critically engage and resist industrial capitalism and its wide-reaching destruction on the other. Though the means of production of vegan foodstuffs certainly deserves scrutiny and vegans should be concerned about the intersecting oppressions that food production currently entails, this does not undercut the need for V_A as an ethical response to violence against animals. One can *both* forgo environmentally destructive products that may also involve human servitude or exploitation and also refrain from consuming animal bodies. Though veganism is one way among many of engaging ethically with animals, it does not follow that those who are well positioned to act should not do all they can to further their goal of ending violence when those actions don't compromise achieving comparable morally worthy ends.[11]

Human beings are always entangled in violence and killing, but there are different responses to these complex entanglements. While there is too much violence globally, much of it, like violence against animals, is systematic. Individual choices and actions in the face of such mass destruction may not appear to do much immediately to stop the violence, but this recognition shouldn't obscure responsibilities to avoid causing harm. Individual animals are victims of mass killing and we humans are, arguably, complicit in their suffering and exploitation.

---

10. For a more thorough discussion of this argument and its weaknesses, see Warkentin (2012).

11. It is also important to note that eating plant-based foods is not a deprivation and is healthier and delicious!

Complicity and Impotence

Just how responsible we are in causing suffering and harm to other animals when we consume their bodies produced in the industrialized system and what difference we might make as individuals, one way or the other, has increasingly been the topic of discussion.[12] One common argument used to reject veganism is that individually, we can't make any difference at all. If V_I rests on a category mistake, V_A rests on a goal that is impossible for anyone to reach.

Consider the context in which one might refrain from eating animals—it is usually when one decides to walk past the animal's carcass in the frozen-food case or the deli section of the local supermarket or butcher where animals are already dead. When one orders a chicken burrito at Chipotle rather than the tofu sofrita, the chickens aren't slaughtered-to-order, so buying the tofu doesn't save any particular chicken's life. The animal bodies in the supermarket or at the restaurant were killed days or weeks before any consumer even thought about purchasing them. Not purchasing a chicken burrito at the particular moment you are at Chipotle would have absolutely no effect either way on whether chickens suffer and die in food production, so refraining from purchasing that chicken prevents no harm.[13] Agribusiness seems to be too massive to respond to the behavior of individual consumers.[14]

Further, consider the case of leftovers. Suppose your housemate brings home leftover chicken. The chicken is already dead and already cooked. Your housemate does not want to eat the rest of his meal. If you do not eat it, the meal will be thrown away. It's hard to see how your eating this leftover chicken could, in any way, add to the harm and misery of factory-farmed chickens, or fail to prevent further violence against such sentient beings. Since animals suffer no matter what you do, why not order or eat the chicken? With regard to individual actions of individual consumers, it seems veganism, particularly as an aspiration, is useless as a response to violence, exploitation, and domination.

---

12. See also Kagan (2011).

13. There is a related debate about group complicity, that is the immorality of even purchasing vegan food in restaurants or stores where animal products are also sold that we can't take up here. See Martin (2015).

14. For a nice overview of this causal impotence objection, see Bass (n.d.)

This contradicts the claim some vegans make that by forgoing products of cruelty, they save 95 animals every year.[15] Presumably they mean that an estimated 95 animals will not be born to become someone's meal. But these 95 indeterminate individuals aren't benefited by not being brought into existence. If you cannot harm or save a non-existent being, it seems no one is saved by not eating animals. In addition, as the number of people who opt out of animal consumption continues to increase, so too does the number of animals killed for food globally; there doesn't even appear to be a correlation between the overall number of animals killed for food and actual individual decisions to abstain from consuming animals. Given all of this, it seems that V_A is illusory. To do what one can to refrain from consuming products that require the suffering and death of other animals amounts to doing nothing to save animals who are suffering and dying.

But how is it possible that individual actions have no impact when it is clear that if everybody abstained, it would make a very large difference? Of course, animals would be spared lives of misery if people ceased consuming animal products, yet it appears that no particular animals would be spared lives of misery if *I as an individual* ceased consuming animal products. As Shelly Kagan puts it:

> it seems to be the case that whether or not I buy a chicken makes no difference at all to how many chickens are ordered by the store—and thus no difference in the lives of any chickens. To be sure, when hundreds of thousands of us each buy a chicken this week, this does make a difference—for if several hundred thousand fewer chickens were sold this week, the chicken industry would dramatically reduce the number of chickens it tortures. Thus the overall result of everyone's buying chickens is bad. But for all that, it seems true that it makes no difference at all whether or not I buy a chicken; even if I don't buy one, the results are no better (Kagan, 2011, p. 110–111).

But how can I make no difference if together we can make a difference? If collective action will have causal impact, then at least *some* individual instances must have causal impact. Collective action is not a particularly mysterious metaphysical category; it is some combination of individual actions that can

---

15. See for example the book, *Ninety-Five: Meeting America's Farmed Animals in Stories and Photographs*, edited and published by No Voice Unheard, 2010. See also PETA (2010).

have a variety of impacts. In some instances a perceptible harmful result emerges from actions that lead to seemingly imperceptible harms. Usually, analyses of these types of situations reveal that though seemingly imperceptible, there is nonetheless some very small impact that, when combined with the very small impacts of other consumers, results in harm. In the cases we are talking about, this seems an unsatisfying way of answering the question, given that the animals are already dead before I even formulate an intention to purchase their bodies. Eating or not eating a dead animal doesn't causally contribute to any animal's death.

But it may be that my action serves as a "trigger" or "threshold."[16] Suppose that the butcher only makes a call to order more chickens when the 100th chicken breast is purchased or the poultry industry only reduces production when a threshold of 10,000 people stop purchasing chicken. It may seem that if you are not the one who purchases the 100th chicken breast or are not the 10,000th person who gave up chicken products, your refraining from such purchases makes no difference. However, your refraining affects the timing of slaughter or the cessation of slaughter. This is an impact, even if it is not a direct impact on any particular individual. So buying or not buying animal bodies *does* make a difference. Further, no matter what the causal impact of your refraining from consuming animal products, what is certain is that your not going vegan is practically certain to delay any threshold event happening and therefore practically certain to result in excess animal suffering (Norcross, 2004).

Recognizing one's complicity in a system of violence and deciding to stand against it by refusing, as far as is possible, to participate in or directly benefit from that system also, importantly, has effects on others. Many who work

---

16. Kagan describes a triggering event in this way:

Presumably it works something like this: there are, perhaps, 25 chickens in a given crate of chickens. So the butcher looks to see when 25 chickens have been sold, so as to order 25 more. (Perhaps he starts the day with 30 chickens, and when he gets down to only 5 left, he orders another 25—so as never to run out. But he must throw away the excess chickens at the end of the day before they spoil, so he cannot simply start out with thousands of chickens and pay no attention at all to how many are sold.)

Here, then, it makes no difference to the butcher whether 7, 13, or 23 chickens have been sold. But when 25 have been sold this triggers the call to the chicken farm, and 25 more chickens are killed, and another 25 eggs are hatched to be raised and tortured. Thus, as a first approximation, we can say that only the 25th purchaser of a chicken makes a difference. It is this purchase that triggers the reaction from the butcher, this purchase that results in more chicken suffering.

toward veganism influence others to do so, and they in turn can influence others, and so on. This kind of role modeling may be understood as a species of the broader phenomenon of *social contagion* in which an action of a particular type makes another action of that type more likely. Thus veganism increases the probability that others will become vegan, which increases the probability that the collective action of the aggregate more quickly brings about a reduction in the number of animals produced for food and other consumer goods, decreasing animal suffering and bringing about a decrease in violence, exploitation, and domination (Almassi, 2011).

In contrast, private actions like eating the leftover chicken when no one else is around (or will ever witness or even find out about it) could increase the chance that one may, in the future, eat more chicken. An internal, private permission is generated and it may expand to other, less private, contexts. Veganism urges us to conceptualize chicken or pig bodies, for example, as "not food," much the way we in the United States think of dog bodies as "not food." As people begin to view the corpses of others as inedible, the probability that they will want to consume "leftover" bodies is lowered. Someone aspiring to be the kind of person who acts to minimize suffering and oppression, wherever and whenever they can, will thus adopt strategies that will stabilize their ability to act on their aspiration and refrain from consuming animal products even in private.[17]

Conclusion

People are looking for alternatives to the systemic, industrialized violence animals suffer in order to become dinner. Though veganism remains an empowering response to this violence, vegans need to remain realistic about the ethical entanglements that accompany life in consumer culture. To believe, as some do, that veganism is an identity or lifestyle that gives one "clean hands" is to believe a myth. In contexts like ours, veganism can only be an aspiration. But even as an aspiration, veganism can make a difference in changing systematic cruelty and domination.

---

17. Interestingly, when considering that role-modeling behavior can have both positive and negative aspects and recognize that some "negatively contagious" actions (so-called "backfire" role-modeling) can affect others' behavior such that it increases the probability that an observer will engage in behaviors *opposite* to the role-modeler, we have further evidence against V_I. If advocates of V_I are perceived as preachy, self-righteous zealots (the "negative contagion"), then the effect of V_I may very well be to push non-vegans *away* from veganism and toward meat consumption.

Abstaining from the use of all animal products is virtually impossible for most consumers in industrialized societies. Coming to think of veganism as an aspiration is coming to terms with the complicated impacts of our choices and relationships with nonhuman animals and the environment. Because it is non-idealized, V_A forges a particularly empowering and grounded form of individual political commitment, fostering a deeper understanding of intersecting injustices and oppressions. In our experience, discussing veganism not as an identity or lifestyle but as an aspiration allows for meaningful discussions about the ways the objectification and commodification of sentient beings are morally problematic. Relatedly, in avoiding the rhetoric of moral purity or superiority, V_A increases the likelihood that non-vegans will be open to embracing the nonviolence that grounds veganism. Recognizing the kind of impact aspiring to veganism can have may strengthen one's ability to respond to the system of violence and improve the lives of all beings.

References

Adelman, Jacob (2009) "Mobile Slaughterhouse Assists in Trend to Locally Killed Meat." Retrieved May 12, 2014, from www.gosanangelo.com/news/2009/jul/25/mobile-slaughterhouse-assists-in-trend-to-killed/.

Almassi, Ben (2011) "The Consequences of Individual Consumption: A Defence of Threshold Arguments for Vegetarianism and Consumer Ethics," *Journal of Applied Philosophy*, 28(4), 396–411.

Applestone, Joshua (2011) "Q&A with Joshua and Jessica Applestone," *The Butcher's Guide to Well-Raised Meat*, Amazon.com Editorial review. Retrieved April 14, 2014, from http://www.amazon.com/The-Butchers-Guide-Well-Raised-Meat-Poultry/dp/0307716627.

Bass, Robert (n.d.) "What Can One Person Do? Causal Impotence and Dietary Choice." Unpublished manuscript.

Bohanec, Hope (2013) *The Ultimate Betrayal: Is There Happy Meat?* iUniverse.

Bohanec, Hope (2014) "Factory Farming vs. Alternative Farming: The Humane Hoax," Free from Harm. Retrieved April 14, 2014, from http://freefromharm.org/animal-products-and-ethics/factory-farming-alternative-farming/.

Dominick, Brian A., (1997) "Animal Liberation and Social Revolution: A Vegan Perspective on Anarchism or An Anarchist Perspective on Veganism." Retrieved April 9, 2015, from http://theanarchistlibrary.org/library/brian-a-dominick-animal-liberation-and-social-revolution.

FFH, (2011) "59 Billion Land and Sea Animals Killed for Food in the US in 2009," Free from Harm. Retrieved April 14, 2014, from http://freefromharm.org/farm-animal-welfare/59-billion- land-and-sea-animals-killed-for-food-in-the-us-in-2009/.

Gillespie, Kathryn (2011) "How Happy Is Your Meat?: Confronting (Dis)connected-ness in the 'Alternative' Meat Industry," *The Brock Review*, 12(1), pp. 100–128.

Gruen, Lori (2011) *Ethics and Animals: An Introduction*. Cambridge: Cambridge University Press.

Gruen, Lori (2014) "Facing Death and Practicing Grief" in *Ecofeminism: Feminist Intersections with Other Animals and the Earth*, Carol J. Adams and Lori Gruen, eds., New York: Bloomsbury Press, pp. 127–141.

Gruen, Lori (2015) *Entangled Empathy: An Alternative Ethic for Our Relationships with Animals*. New York: Lantern Press.

Gruen, Lori, and Clement Loo (2014) "Climate Change and Food Justice" in *Canned Heat: Ethics and Politics of Climate Change*, M. Di Paola & G. Pelligrino, eds., London: Routledge.pp. 179–192.

Hawthorne, Mark (2013) "The Problem with Palm Oil," *Veg News*. Retrieved April 14, 2014, from http://vegnews.com/articles/page.do?pageId=5795&catId=1.

Jenkins, S., and V. Stănescu (2014) "One Struggle" in *Defining Critical Animal Studies*, Anthony J. Nocella II, John Sorenson, Kim Socha, and Atsuko Matsuoka, eds., NY: Peter Lang Publishing, pp. 74–85.

Kagan, Shelley (2011) "Do I Make a Difference?," *Philosophy & Public Affairs*, 39(2), pp. 105–141.

Kheel, Marti (2004) "Vegetarianism and Ecofeminism: Toppling Patriarchy with a Fork." In *Food for Thought: The Debate Over Eating Meat*, Steve F. Sapontzis, ed., NY: Prometheus Press, pp. 327–341.

Martin, Adrienne (2015) "Consumer Complicity in Factory Farming" in *Philosophy Comes to Dinner: Arguments on the Ethics of Eating*, Andrew Chignell, Terence Cuneo, and Matthew Halteman, eds., NY: Routledge.

McWilliams, James (2011) "Killing What You Eat: The Dark Side of Compassionate Carnivorism," Freakonomics. Retrieved April 27, 2014, from http://freakonomics.com/2011/09/20/killing-what-you-eat-the-dark-side-of-compassionate-carnivorism/.

Norcross, Alisdair (2004) "Puppies, Pigs, and People: Eating Meat and Marginal Cases." *Philosophical Perspectives*, 18(1), 229–245.

PETA (2010) "Vegans Save 198 Animals a Year." Retrieved April 27, 2014, from http://www.peta.org/blog/vegans-save-185-animals-year/.

Plumwood, Val (1996) "Being Prey," *Terra Nova: Nature and Culture*, Cambridge, MA: MIT Press, 1(3), pp. 32–44.

Plumwood, Val (2000) "Integrating Ethical frameworks for Animals, Humans, and Nature: A Critical Feminist Eco-Socialist Analysis," *Ethics and the Environment*, 5(2), pp. 285–322.

Steinfeld et al. (2006) Livestock's Long Shadow: Environmental Issues and Options, Food and Agriculture Organization of the United Nations, http://www.fao.org/docrep/010/a0701e/a0701e00.htm.

U.S. Senate Committee on Agriculture, Nutrition and Forestry (USSCANF) (1997) Animal Waste 105th Congress, 1st Session. "Pollution in America: An Emerging National Problem." Report compiled for Senator Tom Harkin. <http://www.grida. no/geo/GEO/Geo-2-260.hum>.

Warkentin, Traci (2012) "Must Every Animal Studies Scholar Be Vegan?," *Hypatia* 27(2), pp. 8–13.

10 VEGETARIANISM: TOWARD IDEOLOGICAL IMPURITY

Neil Levy

In August 2011, a number of news outlets reported the horror and disgust experienced by a Hindu woman who was served meat on-board an international flight. Though she ate only a little chicken before she realized what it was, the woman was profoundly upset by the accidental violation of her dietary regime. A lifetime of vegetarianism was, according to reports, "undone" by the meal (Berry 2011). Now I have no wish to mock anyone's religion or to minimize the distress that this woman may genuinely feel. I want to turn the spotlight on myself, rather than this woman. I want to highlight the shock of recognition I feel (a shock, I bet, many other vegetarians would also experience) when I consider how I would react in similar circumstances. Like the Hindu woman, I would be upset by having accidentally consumed meat. Like her, I would feel, somehow, *defiled*; like her I would think that the meal threatened to "undo" my adherence to a vegetarian diet. *Unlike* the woman, I accept no overarching metaphysical claims—concerning gods and their wishes, or the nature of the universe—that could justify my horror. Rather, I justify my vegetarianism on certain sorts of ethical grounds concerning animal welfare and the environmental costs of animal agriculture. Why, therefore, do I respond as though I would be somehow defiled by an inadvertent mouthful of meat?

In this chapter, I will argue that vegetarianism often takes on a quasi-religious status for vegetarians. It may be appropriate for vegetarians to be "strict"—to have a no-meat rule, which they apply without exception—because rules can simplify decision-making and promote wanted patterns of behavior, thereby enabling agents better to achieve their goals. But, I will suggest, sacralizing the rule may actually have costs for vegetarians, where these costs are measured by the degree to which they allow us to achieve our goals. We might be better off following rules we self-consciously see as justified (compared to alternatives that would achieve our ethical goals as well, or nearly as well) by nothing more than our commitment to them, rather than rules we see as somehow cutting nature at its joints (to use an all-too-carnivorous metaphor). I will suggest that the adoption of rules grounded in personal commitments, just insofar as they are seen as binding on the individual herself and not as having some kind of universal force, may facilitate making common cause with others who do not (yet) wish to commit to a meat-free diet but who are willing to take concrete steps toward reducing animal suffering and environmental harm. Further, under a variety of circumstances adopting these kinds of rules will better enable at least some vegetarians to stick to a meat-free diet.

In what follows, I make some assumptions, for the most part without pretending to argue for them. I assume the truth of (at minimum) a broad kind of naturalism. I assume, that is, that the universe contains the kinds of facts and properties that feature in good science, as well as any kinds of facts and properties that can be entirely explained by reference to these, and nothing else. This naturalism is meant to be as uncontroversial as possible, merely ruling out appeal to supernatural properties (it therefore is neutral on a variety of controversial philosophical issues, for instance concerning whether properties like "good," "beautiful," "consciousness," and so forth can be *reduced* to more basic facts and properties or merely explained by reference to them). It insists that when we appeal to properties or events that do not feature in science these properties must nevertheless be consistent with our best science. Moreover, those who invoke them bear a burden of proof: these properties or events had better earn their keep in explaining the world, including the human world, and must be explicable. I assume, further, that vegetarianism is justified (to the extent to which it is justified) by reference to concerns with animal welfare and/or the environment. More contentiously, I assume some kind of consequentialism is true, at least with regard to the questions of animal welfare and environmental harms (that's compatible, of course, with thinking that human beings are rights-bearers and therefore rejecting consequentialism as an overarching normative theory). My arguments are not addressed to vegetarians like the Hindu woman mentioned above; rather, they turn on the conflict between the justifications I assume and the aptness to experience emotions that she (but not we) is justified in feeling.[1] Given that we do not justify our vegetarianism by reference to a metaphysics that makes sense of an experience of defilement—that we *reject* such a metaphysics—why do we nevertheless experience emotions akin to hers, and ought we to be concerned about the fact that we do?

In his recent work, Haidt and colleagues (Haidt & Kesebir; Haidt 2012) have identified six alleged "foundations" of morality, where a "foundation" is a dimension along which we implicitly categorize actions and agents; this categorization generates an intuition—roughly, an immediate sense that something is right or wrong—when we perceive or imagine events or objects that fall within the scope encompassed by the foundation. Thus, for instance, if we categorize an action as high on the dimension of "fairness" (to mention one foundation), we are likely to generate an approving intuition. The six foundations Haidt and colleagues identify are:

1. Care/Harm
2. Fairness/Cheating
3. Liberty/Oppression
4. Loyalty/Betrayal
5. Authority/Subversion
6. Sanctity/Degradation

This list is not meant to be exhaustive; Haidt and colleagues countenance the possibility that there may be more foundations. However, they claim that these six foundations, perhaps when supplemented with several more, explain and describe the moral judgments of most human societies.

These foundations are supposed to be innate, having an evolutionary explanation. For instance, the concern about fairness may have its origins in reciprocal altruism—behavior that boosts organisms' inclusive fitness by trading aid—while concerns with sanctity might have an original basis in adaptations designed for pathogen avoidance. Because they are innate,

---

1. The restriction of my argument to those who justify vegetarianism on certain sorts of naturalistically respectable ethical grounds is far from trivial. There are prominent advocates of vegetarianism who accept metaphysical frameworks that might justify a sense of defilement; for instance, those like Regan (1983) who hold that animals have moral rights and perhaps those who defend vegetarianism on virtue-ethical grounds, such as Hursthouse (2006), though it is not clear to me that virtue-ethical vegetarians must maintain that eating meat is defiling.

human beings come prepared to categorize the world in their terms: though they may be culturally modulated and sometimes cultures may even come near to entirely suppressing one or more foundations, the grain of human nature ensures that it is easier to build moral systems on these foundations than against them.

Controversially, Haidt argues that moral foundations theory helps to explain and, to some extent, dissolve, political controversies. He claims that "liberals"—using that word in its US meaning, to refer, very roughly, to social democrats (rather than to refer to people who ascribe to liberal political philosophy)—have a morality that is based almost exclusively on concerns about harm and fairness alone, whereas conservatives have a morality that is sensitive to all six moral foundations. Liberals accept a moral outlook that is deaf to concerns with sanctity, authority, and loyalty.

If Haidt is right, the Hindu woman mentioned at the beginning of this chapter, like adherents to traditional overarching conceptions of the good across the world, picks out normative properties by reference to a larger set of foundations than does the average US liberal. She perceives some actions or objects as morally loaded in ways that the liberal would not and sees some as morally loaded that the liberal would see as morally neutral. She may well accept a social ontology that makes sense of notions of hierarchy, and therefore authority and its subversion (insofar as she accepts the traditional Hindu caste system, she is committed to accepting these claims). More relevantly to our concerns, she clearly accepts claims concerning sanctity and degradation. These claims are explicit in what she says about why she is a vegetarian ("I gave meat to the god"). Now, to regard something as sacred, in this way, is to make certain claims about an extra-personal moral order: it is to claim that the nature of things dictates that certain practices or things have a categorical status that makes them to-be-done or to-be-avoided; adherence to the practices dictated does not rest on convention (as might adherence to the loyalty/betrayal foundation, or the authority/subversion foundation, either or both of which might be founded on tradition) but on the nature of things. They are *intrinsically* to-be-done or avoided. Even if they owe their justification to divine command, the command invests them with a sacred status in themselves—it is not merely that we are commanded to do thus and so; rather, *because* of the command that we do thus and so, it is intrinsically right that we do thus and so. In brief: sanctification of a practice or action entails the postulation of properties that sit uneasily within the broad naturalism I am here assuming. Any such properties require justification before we have the right to invoke them.

Though I will not attempt to argue for the claim, I am going to assume, further, that the sanctification of food *cannot* be justified within the broadly naturalist framework within which I am working (this assumption is not innocuous: a Wittgensteinian view of religious commitments was once influential, on which religious belief did not necessarily entail robust realism about God; perhaps something analogous could underwrite sanctification. Whatever the merits of the Wittgensteinian (or otherwise non-realist) positions in the philosophy of religion, I suspect that ordinary believers are, and take themselves to be, committed to robust realism about God, and that a non-realist position is difficult to maintain outside the study; sanctification doesn't entail a non-naturalistic metaphysics, but it constitutes strong psychological pressure toward one, I suggest. Claims about the evolutionary basis of the sanctification foundation makes this plausible, I suggest, but it is a hypothesis in need of confirmation). The assumed framework is intended to be neutral on the justifiability of moral realism, but while there are plausible moral realisms compatible with the broadly naturalistic framework (see, for instance, Boyd 1988), I claim that there are no such plausible naturalistic accounts that could justify attributing to foodstuffs the kind of properties invoked by the sanctification/degradation foundation. I therefore go further than Haidt, in claiming that not only is it true that liberals are (relatively) blind to foundations of morality other than the care/harm and fairness/cheating foundation, but also that they are *rightly* blind to the sanctification/degradation foundation. Liberals need not be blind to the wisdom of tradition, nor need they deny that certain individuals should command our respect, so liberals need not be blind to notions of loyalty and authority. However, the claim that something is sanctified is an ontological extravagance, and ought to be avoided by liberals.

My guilty recognition of the Hindu woman's distress indicates, however, that I have not entirely succeeded in avoiding the sanctification of my own dietary rules. I am sure that I am not alone in this: neither in having sanctified my vegetarianism nor in having done so despite rejecting the metaphysics that could justify doing so. My vegetarianism has become *moralized*, in the terminology of Rozin, Markwith, and Stoess (1997), which is to say that it has implicitly come to be seen by me as *intrinsically* valuable. Rozin et al. found that those people who avoid meat for moral reasons tend to moralize meat avoidance, whereas those who avoid it for health reasons do not. Just as Haidt would predict, moralized vegetarianism is associated with disgust at meat eating by Rozin et al.; disgust is the moral emotion preferentially triggered by violations of the sanctified (Graham et al. 2013). For liberal naturalists like me, this sanctification is in conflict with the framework we accept.

Why do we nevertheless tend to sanctify vegetarianism? Perhaps Haidt is right, and the sanctification/degradation foundation is innate, or, at any rate, easily triggered in us. In line with this claim, Haidt cites several pieces of evidence for a conflict between liberals' considered judgments with regard to violations of norms of purity (as well as norms concerned with authority and loyalty) and their implicit attitudes. For instance, Skitka et al. 2002 found that cognitive load altered liberals' judgments to more closely resemble those of conservatives, suggesting that liberals' judgments are produced by an effortful overriding of their intuitive responses. If this is correct—if it takes effort to avoid perceiving moral violations by reference to the sanctification foundation, whether we explicitly accept it or not—it is unsurprising that we find ourselves tending to sanctify our deeply held moral convictions.

Should we be concerned about our sanctification of our preferences? Of course, insofar as our sanctifying our preferences is inconsistent with our broader views, there is *some* reason to be concerned. In general, we ought to prefer properly justified and consistent views to those that are inconsistent and which we cannot (therefore) justify. An inability to justify our views leaves us at a disadvantage in debates and open to the charge of hypocrisy. Insofar as inconsistency generates cognitive dissonance, moreover, it is possible that we will not be able to limit its effects: it may contaminate our other beliefs. There is extensive evidence that cognitive dissonance leads people to confabulate attitudes. For instance, the inability to justify to oneself why one has written an essay supporting a view that it is rather unlikely that (antecedently) one actually accepts (say an essay supporting raising tuition, written by a college student) leads people to self-attribute the attitude. In contrast, controls induced to write the essay for an amount of money sufficient to justify doing so do not self-attribute the attitude (see Cooper 2007 for exhaustive review). An inability to justify one's disgust and sense of defilement at the prospect of eating meat might constitute pressure to adopt the sanctity/degradation foundation, and that would have consequences for one's other moral views. Hence, we cannot be sanguine in the face of our propensity to sanctify vegetarianism.

Moreover, and more directly relevantly, the propensity to sanctify vegetarianism may actually be an obstacle to the achievement of the goals at which I am assuming the liberal vegetarian aims (the reduction of unnecessary suffering and the avoidance of harms to the environment). One important reason is this: regarding vegetarianism as a matter of purity entails, as the report concerning the Hindu woman with which we started clearly indicates, that a transgression or a violation is never a small matter. For the Hindu woman, her accidental consumption of meat threatened to "undo" a lifetime of vegetarianism. Seeing a behavior as a matter of purity encourages an all-or-nothing viewpoint with regard to violations. Though purity is a matter of degree, there is an asymmetry concerning the way in which it increases and at least some of the ways in which it is reduced. There is no way for someone to achieve a high degree of purity all at once (short of a miracle); instead, one accumulates purity gradually, through living in the right way. But it is quite possible for one to become very impure all at once: defilement returns one to degree zero of purity. If one sees vegetarianism as a matter of purity, falling off the wagon entails no longer being a vegetarian at all. And that may lead people to see a single violation as entailing that there is no point in avoiding meat any longer, at least for some period of time.

This is because accumulating purity takes so much time: if I avoid meat (or sex, or drugs, or what have you) for two days, I don't count as particularly pure. Rather, I am pure if I avoid a proscribed activity for some very significant period of time. Though one violation sets the scale back to zero, the longer one has abstained the higher one's "spiritual" value. This fact, the fact that achieving purity requires a considerable investment of time, entails that the marginal value of a short period of time after a violation, which sets the value back to zero, is tiny. So if I have fallen off the wagon, I might as well go—or eat—the whole hog. I won't get any less pure; you can't go lower than zero. And abstaining from eating meat for the next twenty-four hours won't raise my purity level noticeably. Indeed, this kind of logic predicts that people may not feel it worthwhile to return to vegetarianism *at all* after a transgression. Since it would take a very long time to return to one's former level of purity, and since the months, years, even decades, of investment have been dissipated in one moment, many people who adopt this way of thinking lack the capacity to motivate themselves to start again. The disparity between their former purity and their current degraded state, together with the contemplation of the enormous investment of time (if not effort—whether vegetarianism is onerous differs from person to person) required to return to their former state, saps their motivation. This, I suggest, explains a common narrative: many people have told me that they were vegetarians for many years, but they are no longer, despite the fact that they remain convinced of the moral and/or environmental case for vegetarianism. Conceptualizing vegetarianism within the sanctity/degradation framework predicts difficulty in remaining a vegetarian once one falls off the wagon, whether one does it purposefully (as those who find meat tempting may do) or accidentally, like the Hindu woman with whom we started (in line with this latter suggestion, it is interesting that unlike moral anger, negative evaluations of sanctity violations do not track whether the violation was intentional or not: see Young and Saxe (2011)).

Conceptualizing vegetarianism in terms of sanctity might also lead to the division of the world into in-groups (proper vegetarians) and out-groups (everyone else). Purity is an ideal that is ever-receding: just as in religion people often seem to attempt to outdo one another on the purity front, by raising the standards for what counts as genuinely pure and then denouncing those who are less strict, less fundamentalist than themselves as no better than heretics or atheists (in politics, think of the RINO slur—Republican in name only—often directed by members of the Tea Party at those slightly less right-wing than they are), so in any arena conceptualized in terms of purity, coalition-building becomes difficult. Those seen as less pure than oneself are regarded not as potential allies, but as traitors to the cause. It is very likely that forces akin to this one are at work within strands of vegetarianism: think of the contempt that some vegans show for mere lacto-ovo vegetarians, a contempt that some of the latter might show for those who try to avoid meat (perhaps abstaining from meat one day a week). Again, insofar as our goal is to reduce animal suffering and harm to the environment, we should see one another as allies; sanctification undermines our capacity to do so.

The goal of the vegetarians who are the subject of this chapter is to reduce unnecessary suffering and harms to the environment. Insofar as conceptualizing vegetarianism as a matter of purity may actually lead to more meat eating, rather than less, we have reasons to avoid such sanctification. Insofar as one wants to achieve these ethical goals, it might be preferable to adopt a less strict regime, one that might be less apt to be moralized: perhaps something like "*avoid meat* (unless one is reasonably sure that it is sourced in a manner that doesn't raise ethical or environmental concerns)." This far more relaxed rule would, if followed consistently, achieve the goals of the liberal vegetarian about as well as a much more strict rule, and may avoid the costs associated with cognitive dissonance, ideological purity, and a propensity to see a single violation as a catastrophe, all of which may arise from adherence to a rule that is more apt to be sanctified.[2]

---

2. It is an empirical issue whether the costs of disgust are larger than the benefits. As Bob Fischer has pointed out to me, disgust may be triggered by exposure to the ways in which meat is produced—practices that it is natural to describe as disgusting—and such disgust might motivate vegetarianism; any disposition toward sacralization that follows might be a price worth paying. A great deal turns on how powerful the mechanisms of cognitive dissonance are, and what their broader effects on the cognitive economy can be expected to be, as well as how common lapses that return the sacralizing vegetarian to omnivorism actually are.

However, there may be countervailing pressures: there are reasons why one might want to adhere to a strict rule rather than one that is more relaxed. First of all, fuzzy rules are harder to follow than stricter rules. As George Ainslie (2001) has emphasized, the extent to which rules must be interpreted in order to be applied predicts their liability to encourage self-deception. Rules like "minimize the consumption of unethically sourced meat" are not clearly action guiding. What counts as unethically sourced (should I accept the waiter's assurance that the pig was slaughtered humanely? Should I assume that it was farmed intensively? What were the environmental impacts of its production?)? What kinds of actions are most likely to minimize the consumption of such meat? Might my abstaining now serve as a good example to others or, conversely, will they think that I am being sanctimonious and actually be *less* likely to cut down their own meat consumption in future as a result of my actions? The latter thought might especially lend itself to self-deception, especially if I am tempted to eat meat; so might the awareness of this temptation, coupled with fuzzy guidelines, by encouraging me to think that consumption *now* might enable me to better abstain in future.[3]

All of these points are of course familiar from debates within consequentialist ethics; these are among the kinds of considerations that lend support to a rule-consequentialism, which advises following a set of rules (more or less) without exception, rather than attempting to calculate the expected utility of actions from moment to moment. Further, fuzzy rules are costly to implement, inasmuch as they are demanding of time and resources. Effortful processing produces cognitive load, and a concomitant decline in performance on tasks that are demanding. Worse, cognitive load decreases self-control (Levy 2011); if the person finds meat tempting, then using resource-intensive rules may increase the likelihood of giving in to temptation. Finally, fuzzy rules may also have perverse effects. Human beings exercise—and lose—self-control, as Ainslie suggests, by regarding their past actions as predictive of how they will behave in the future; the more so because the costs and benefits of self-control are often cumulative (the marginal benefit of not smoking one cigarette at time *t* is trivial; the cumulative benefits of a policy of not smoking may be very large). For this reason, fuzzy rules may tend to

---

3. Eric Schwitzgebel and colleagues have found that philosophers with a specialization in ethics don't seem to behave any better than colleagues in other fields. One possible explanation for this fact is that expertise in moral reasoning enables more sophisticated rationalization (see Rust and Schwitzgebel 2014). If that explanation is (partially) correct, philosophers might be especially vulnerable to misapplying fuzzy rules.

gradually narrow in scope, as my failure to abstain in a particular circumstance (it's a special occasion; I've had a hard day; it would be anti-social to insist) predicts a failure to abstain in similar circumstances in the future. At the extreme this may lead to the formation of what Ainslie (2001) calls a "lapse district"; a set of circumstances in which past lapses so strongly predict future lapses that the person no longer sees themselves as being able to exercise self-control in them. A bright line, by preventing such lapses in the first place, may better enable the avoidance of such districts as well as to ensure desired behavior.

In general: because judgment calls are difficult and demanding, and open to abuse, bright lines are often preferable to such calls. For very many people, under many circumstances, it is easier to abstain from eating tempting goods than to eat them in moderation; similarly it is easier—for many people—to be "strict" in one's vegetarianism than to limit meat consumption, even if one's goals of reducing animal suffering and harm to the environment would be achieved about as effectively by a less strict regime, were one able to follow it. But strict regimes have costs as well as benefits: in particular, they lend themselves to sanctification and that might undermine our capacity to achieve the goals toward which we aim, both by alienating potential allies and by leading us to see a single lapse as a catastrophe.

The situation we confront is therefore this: Strict rules lend themselves to sanctification, and such sanctification has costs, where "costs" are measured in their effectiveness at achieving the goals I here assume motivate the vegetarian. But fuzzy rules also have costs, measured in the same currency. Strict and fuzzy, used as I use the terms here, are exhaustive classifications. However we choose, we choose options that will have some costs. We should therefore attempt to choose the option that best enables us to achieve our ethical goals. Since fuzzy rules and precise rules, if followed successfully, achieve our ethical goals about as well, we should choose which we follow on the basis of which has the lowest costs.

On balance, I suggest, strict rules are likely to be less costly than fuzzy ones. Unlike some rules that are apparently strict, the rule "do not eat meat" (or alternatives like "do not eat animal products"; "do not eat meat or diary products") are relatively easy to apply, in part because there are broadly shared understandings of the scope of these rules, and therefore they are cognitively cheap to apply and do not easily lend themselves to self-deception. The main risk of strict rules is that such rules lend themselves to sanctification, which threatens, in turn, to cause alienation of potential allies, the perception that a single lapse is catastrophic and perhaps, via the mechanisms of cognitive dissonance, the confabulation of attitudes we ought to avoid. I therefore suggest that vegetarians aim to adopt strict rules while guarding carefully against sanctification of these rules.

That, of course, is far easier said than done. How do we avoid sanctification? I know of no easy method that is guaranteed to succeed. I suggest that the best approach may be to adopt what I will call *existentialist* rules; rules that are explicitly and self-consciously adopted (from among those that achieve our ethical goals) through what are seen as acts of choice and justified *because* they are chosen. It is this foundation in an act of choice that justified the label "existentialist," though unlike the acts of choice associated with Sartre, these actions are not to be understood as creating values *ex nihilo*; rather, only those rules that promote our ethical goals are available for choice. That one such rule is justified, rather than another among this set, is then to be understood as justified by our choice. Rather than seeing these rules as binding because they cut nature at its joints, we should see them as binding on us (only) because we have chosen them, and therefore as not binding on others unless they, too, undertake an act of choosing to be bound; an act that is not necessitated or required by anything beyond that act of choice.

I am far from sure that this approach will succeed.[4] The central idea is to guard against sanctification by recognizing that the rules are not uniquely justified; perhaps other approaches might succeed as well or better at achieving this end. In any case, it seems likely to have benefits. Adopting existentialist rules will allow us to hold ourselves to stricter standards without seeing those with different rules, but who are making some kind of effort to promote the goals of a reduction of animal suffering and environmental protection, as impure. It allows us to respond to ourselves with different, and much more condemning, reactive attitudes if we deliberately flout our rules, without condemning ourselves if we stray accidentally. A propensity to have such responses only in these circumstances would play a role in helping us to avoid deliberate infringements, without tempting us to reject potential allies as too impure.

---

4. I am also far from sure that this approach is best for everyone. Some people find themselves, willingly or not, in the role of exemplars; for them, a single violation may leave them open to a charge of hypocrisy (justified or not) and may undermine their public standing. Such people may have reason to adopt stricter rules. To the extent to which lapses are likely even for some of these people, however, they might do better both in adopting and advocating less strict rules: the advocacy is important, inasmuch as it preempts the charge of hypocrisy.

The way in which the adoption of an existentialist rule for oneself combined with the avoidance of sanctification facilitates making common cause with people who are, shall we say, less vegetarian than we are is among the primary benefits it promises. Those who adopt such rules will see things like "meat-free Mondays" as genuine goods, not as feel-good movements to be dismissed contemptuously or condescendingly. It also produces a willingness to be open to revising our rules, in the face of new discoveries. There are a range of possible findings and innovations that might justify the adoption of new rules. One possible innovation concerns the possibility of vat-grown meat (Coghlan 2011). Obviously such meat does not—directly at any rate—involve animal suffering. If it can be mass-produced, its environmental costs might be acceptably low. If this occurs, we might wish to rethink our dietary rules. Another possibility concerns the development of animals who cannot suffer (Shriver 2009); again, animal welfare and environmental concerns might favor promoting consumption over the alternative of remaining meat-free. Independently of such high-tech developments, changes in land use and practices might give us grounds for changing our dietary rules. We may discover, sooner rather than later, that we should sacrifice (direct) animal welfare concerns in return for environmental benefits. This could occur through the consumption of livestock farmed on marginal land unsuited for agriculture or through the consumption of game. Though there may be significant animal welfare costs to these patterns of consumption, these costs could well be offset by their environmental consequences (especially when we bear in mind that the adverse effects of environmental harms can befall non-human animals just as much as humans: the costs of climate change, for instance, are likely to be borne by very many species).

We must not be *too* ready to revise our rules. We lose the benefits of strict rules if we are too ready to revise them. But refusing sanctification is refusing absolute rigidity, and that, on balance, is a good thing. It entails adopting a less absolutist vegetarianism, one more accepting of a variety of ways of pursuing the goals we seek and more accepting of a variety of people, not all of whom are prepared, or should be expected, to adopt a no-meat policy. It entails an openness to abandoning vegetarianism altogether, if an alternative regime is clearly better suited to achieving our goals. Vegetarianism reconceived along the lines proposed here is not pure (nor is it all that simple). But purity is an ideal we should gladly sacrifice.[5]

5. This chapter has benefited immensely from helpful comments from Ben Bramble and Bob Fischer.

References

Ainslie G. 2001. *Breakdown of Will.* Cambridge: Cambridge University Press.

Berry, M. 2011. Vegetarian's Horror at Meaty In-Flight Meal. *The Age*, August 24.

Boyd R. 1988. How to Be a Moral Realist. In *Essays on Moral Realism*, ed. G. Sayre-McCord, 181–228. Ithaca, NY: Cornell University Press.

Coghlan, A. 2011. Meat without Slaughter. *Scientific American* 211/2828: 8.

Cooper, Joel. 2007. *Cognitive Dissonance: Fifty Years of a Classic Theory.* Los Angeles: Sage Publications.

Graham, J., & Haidt, J., Koleva, S., Motyl, M., Iyer, R., Wojcik, S., & Ditto, P. H. 2013. Moral Foundations Theory: The Pragmatic Validity of Moral Pluralism. *Advances in Experimental Social Psychology* 47: 55–130.

Haidt, J. 2012. *The Righteous Mind: Why Good People Are Divided by Politics and Religion.* New York: Pantheon.

Haidt, J., & Kesebir, S. 2010. Morality. In *Handbook of Social Psychology, 5th Edition*, ed. S. Fiske, D. Gilbert, & G. Lindzey, 797–832. Hoboken, NJ: Wiley.

Hursthouse, R. 2006. Applying Virtue Ethics to Our Treatment of the Other Animals. In *The Practice of Virtue*, ed. Jennifer Welchman, 136–155. Indianapolis, IN: Hackett.

Levy, N. 2011. Resisting "Weakness of the Will." *Philosophy and Phenomenological Research* 82: 135–155.

Regan, T. 1983. *The Case for Animal Rights.* Berkeley: University of California Press.

Rozin, P., Markwith, M., and Stoess, C. 1997. Moralization and Becoming a Vegetarian: The Transformation of Preferences into Values and the Recruitment of Disgust. *Psychological Science* 8: 67–73.

Rust, J. and Schwitzgebel, E. 2014. The Moral Behavior of Ethicists and the Power of Reason. *Advances in Experimental Moral Psychology*, ed. H. Sarkissian and J.C. Wright, 91–108. London: Bloomsbury Academic.

Shriver, A. 2009. Knocking Out Pain in Livestock: Can Technology Succeed Where Morality has Stalled? *Neuroethics* 2: 115–124.

Skitka, L. J., Mullen, E., Griffin, T., Hutchinson, S., & Chamberlin, B. 2002. Dispositions, Scripts, or Motivated Correction? Understanding Ideological Differences in Explanations for Social Problems. *Journal of Personality and Social Psychology* 83: 470–487.

Young, L. & Saxe, R. 2011. When Ignorance Is No Excuse: Different Roles for Intent across Moral Domains. *Cognition* 120: 202–214.

11 AGAINST BLAMING THE BLAMEWORTHY

Bob Fischer

> This book is an attempt to think through, carefully and consistently, the question of how we ought to treat nonhuman animals. In the process it exposes the prejudices that lie behind our present attitudes and behavior. In the chapters that describe what these attitudes mean in practical terms—how animals suffer from the tyranny of human beings—there are passages that will arouse some emotions. These will, I hope, be emotions of anger and outrage, coupled with a determination to do something about the practices described.
>
> —FROM THE ORIGINAL PREFACE TO PETER SINGER'S
> *ANIMAL LIBERATION*

Introduction

Let's assume that it's almost always wrong for people like me to eat meat. That is, let's assume that healthy Westerners who live well above the poverty line should, in virtually all circumstances, abstain from consuming animal flesh. And now let's note the obvious: most don't abstain, which means that most people act wrongly on a regular basis.[1] Moreover, it's implausible that these wrongs are excusable: most people eat meat because everyone else does; because it tastes good; because it's widely available and inexpensive; *not* because they need it to survive; *not* because beef lobbyists are holding guns to their respective heads; *not* because they are, unbeknownst to us, utility monsters whose gustatory preferences genuinely do deserve greater moral consideration than the interests of the animals they eat. Hence, most people who eat meat are blameworthy for doing so.

---

1. Throughout, "people" refers to people like me: healthy, Western, human animals who live well above the poverty line. Also, I should note that, throughout this paper, I'm going to call nonhuman animals "animals." This is only for convenience. In particular, it does not indicate that (a) I take there to be some morally important differences between nonhuman animals (as a gerrymandered class) and humans (as a less gerrymandered, but still somewhat gerrymandered class); (b) that I take no nonhuman animal to be a person; or (c) that I take all humans to be persons.

Nevertheless, we tend not to blame them. In general, we are unfazed by the dietary choices of others—and when they bother us, we rarely express as much. My aim here is to argue that this is as it should be, at least for the foreseeable future.

I won't contend that blame would be ineffective nor that our current practices represent the best way to promote the interests of animals. I have no idea whether these claims are true, and I take no stance on them here. Instead, my line is this:

1. If it would be unreasonable to demand that someone behave in a particular way, then, if she fails to behave that way, we shouldn't blame her for it.
2. It would be unreasonable to demand that someone abstain from eating meat.
3. So, if someone eats meat, we shouldn't blame her for it.

Call this *the argument against blaming the blameworthy*. I won't say much in defense of the argument's first premise. I assume that blame involves the negative reactive attitudes—anger, indignation, resentment, outrage, and so on—being directed toward a (perceived) moral offense. Granted, these emotions aren't always expressed. But if they're absent entirely, then I think you've got *judging to be blameworthy*, just plain *moral disapproval*, or some other response in that neighborhood.[2] Moreover, I follow Gary Watson in thinking that "[the] negative reactive attitudes express a *moral* demand, a demand for reasonable regard" (Watson 2004, p. 229). Given as much, the first premise is practically true by definition.

The argument turns, then, on the second premise. Here's a *précis* of the case for it. I'm a vegetarian because I'm moved by standard arguments for vegetarianism: I think the benefits of meat-eating don't outweigh the costs; I think that to consume animals is to fail to treat them as ends in themselves (and they deserve such treatment); I think eating meat is an expression of callousness. But these sorts of arguments generalize, which is to say that they don't merely cast a shadow on meat-eating, but on countless activities that are part and parcel of living in a modern consumer society: for example, spending money in ways that support exploitative labor practices, or investing energies

---

2. I acknowledge that strong negative reactive attitudes may not exhaust blame: it may have other dimensions, such as the judgment that *a* is responsible for *x*, or that *x* is a mark against *a*'s character. I take no stand on what those other dimensions may be.

in our hobbies instead of the homeless shelter. However, since it would be unreasonable to demand that people make each and every one of these changes, we have to reflect on which, if any, we can insist upon. When we do so, we find that meat-eating doesn't make the cut.

To be clear, my goal here is not to argue that meat-eating is morally permissible. It isn't. Nor am I arguing that we shouldn't encourage people to abstain from consuming animals. We should. I am just interested in whether, of the many possible responses to meat-eating, blame is the appropriate one. In most cases, I think it isn't.

Defending the Argument's Second Premise, Step 1

At issue is whether it's unreasonable to demand that someone abstain from eating meat. I assume that, if the many arguments against meat-eating are *un*-successful, then it certainly isn't. So let's suppose that those arguments *are* successful.[3] What follows?

I think what follows is that much of what we do is morally wrong—not just with respect to animals. The most well-known argument against meat-eating, at least in philosophical circles, is probably the utilitarian one due to Peter Singer (Singer 2009; meat-eating fails to satisfy the expected interests of sentient beings, equally considered), followed closely by the rights-based argument due to Tom Regan (Regan 1983; animals have inherent value, and killing them for food treats them as mere resources). More recently, virtue-based arguments have become more prominent (Hursthouse 2000, 2011; Nobis 2002; Walker 2007; Kriegel 2013; these vary in their details, but are unified by a concern with what compassion requires of us). Moreover, there are a host of arguments based on commonsense moral principles

---

3. Someone might think that this claim is vulnerable to an obvious objection. *Suppose that utilitarian arguments against meat-eating work. Then, utilitarianism is true. If utilitarianism is true, then Kantianism is false. But if Kantianism is false, then Kantian arguments against meat-eating fail. Hence, if some arguments against meat-eating are successful, then others aren't.* There are two ways out. My preference is to reject the assumption that utilitarian and Kantian considerations are in competition with one another. I opt for a view where arguing for a moral position amounts to arguing that it's supported by an overlapping consensus of distinct kinds of moral reasons. However, if we prefer not to reject that assumption, then we can treat the antecedent as shorthand for a more careful claim: namely, that there is a set of arguments against meat-eating having compossible premises, and each of those arguments is sound. (Granted, it matters which set of arguments. For simplicity's sake, though, we can proceed as though it's one that includes an argument based on either utilitarianism, Kantianism, or virtue ethics.)

(Curnutt 1997; Engel 2001; Jordan 2001; Norcross 2004; Rachels 2004; DeGrazia 2009).[4]

It isn't contentious that some of these arguments generalize: the utilitarian argument is often criticized on just this point. But there is nothing special about that one. If the arguments against meat-eating work, then we cannot justify harming others, or violating their rights, or being callous toward their needs, simply because life is more pleasant if we do so. But we can harm others by ignoring them; we can violate their rights by participating in systems that exploit them; we can be callous toward them by pretending that the trivial interests of those near and dear outweigh the most pressing interests of those distant strangers. And these things we do—to animals, to humans—without thinking, day in and day out, in countless small ways. We drive when we could walk, which is worse for us, for the environment, and for the creatures that we are bound to hit with enough such excursions. We get the $10 bottle of wine instead of the $8 bottle (or the $5 bottle instead of the box), though the difference will be imperceptible and the extra dollars could go to the shelter. We forget to vote despite the signs, we fail to protest, we tolerate systemic injustice. We spend lavishly on our families; we buy new when used would be as good; we opt for homes in suburbs instead of high rises, ignoring the ills of suburban sprawl. In general, our decisions serve our personal interests, however inconsequential they may be. We are complicit in great evil.

And if the arguments against meat-eating are successful, we're short on excuses. Consider, for example, an appeal to an individual's causal impotence, which might seem to mitigate moral responsibility. However, an individual's buying and eating a chicken breast has no bearing on the number of chickens killed, as her purchase is noise in the grocery store's inventory. If the arguments work, then this must not matter. Nor is ignorance an excuse: surely some people are unaware of the misery in concentrated animal feeding operations (CAFOs) and slaughterhouses; if the arguments condemn these people, then what matters is the ease of access to relevant information, not whether we actually have it. (It's irresponsible not to know.) Finally, having to make minor sacrifices is no excuse: if these arguments work, then we might have to violate various social norms, revise much-loved traditions, learn new

---

4. Additionally, there are arguments that are based on human interests: for example, ones driven by concerns about distributive justice (Rachels 1977; Midgley 1984), environmental responsibility (Singer 2009, Chapter 4), health (Barnard and Kieswer 2004), and feminist critiques of the association between meat and masculinity (Adams 2010). I ignore these arguments here because I don't think that factoring them in would undermine the generalization thesis; moreover, I'm primarily interested in arguments based on the interests of animals.

skills, give up certain conveniences, and so on. But so it goes when your aim is to treat beings as ends in themselves; so it goes when your aim is to be compassionate.

So if the arguments against meat-eating are successful, then we should expect many features of modern life to be morally wrong. Let's call this *the generalization thesis*. Supposing that the arguments against meat-eating *are* successful, how does the generalization thesis bear on whether it's reasonable to demand that people abstain?

Defending the Argument's Second Premise, Step 2

The answer, I think, is that it shifts the burden of proof. I take it to be obvious that it would be unreasonable to demand that people make every change required by the generalized versions of the arguments against eating meat—not because the obligations aren't real, but because the life we can justify is so far from the one we actually live. Someone might concede as much, but maintain that we need no special reason to demand just a *single* change, as opposed to the whole lot. On this view, we don't need a reason that distinguishes one demand over others; we just need to avoid making every demand. This strikes me as akin to suggesting that, even though it would be unreasonable to demand that road users follow every law to the letter, it's reasonable to demand—without explanation—that cyclists come to a complete stop at every stop sign. The response seems utterly arbitrary. (It isn't enough to point out that, in fact, it's the law: that's a reason for the person to comply, not for us to insist that she comply.) So while it may be reasonable to demand that people make some change or other, we need further argument to show that it's reasonable to demand any particular one.

How might such an argument go? I see three strategies. The first is to emphasize the ease of abstaining. The second is to emphasize the severity of the wrong. The third is to limit the context.

The Ease of Abstaining

Someone might observe that it's just so easy to stop eating meat, and hence hard to think that we shouldn't demand abstinence. I gave up meat years ago, and I confess some sympathy with this thought. Still, it can't be right, and this because so many of the changes we could make are equally simple. You don't need to stop going out to eat and donate all the money you'd save; you just need to go to this slightly less expensive place, donating the difference. You don't need to sell your car and bike everywhere to avoid accidental collisions with songbirds and squirrels; you just need to bike when the weather's good and your destination is within three miles. You don't need to go vegan; you just need to cut out eggs. And so on. There are countless small changes that we have moral reason to make, and that we may well be obliged to make if the arguments against meat-eating are successful. And we have means and opportunity to make them. So, we act wrongly when we don't revise our habits, which is to say that we act wrongly all the time. The problem with the objection is that it fails to see meat-eating in this context, and hence fails to see the implications of using ease of change as our guide to those demands that are reasonable.

The Severity of the Wrong

Precise estimates vary, but it's widely acknowledged that billions of land animals and tens of billions of aquatic animals are killed each year for US consumption alone.[5] If it's wrong to kill and consume animals for food, then surely we are grappling with an enormous evil when we consider meat consumption. Does the severity of the wrong make it reasonable to demand abstinence from meat?

First, we should be wary of the way this is framed. No individual is *causally* responsible for national statistics regarding slaughter. (Indeed, when we consider the behavior of a single consumer, it isn't clear that she's causally responsible for *any* animal slaughter. For further discussion, see the essays by Driver, Budolfson, and Littlejohn in this volume.) Second, whatever our account of complicity, it shouldn't entail that an individual is morally responsible for national statistics regarding slaughter. Whatever responsibility attaches to the individual, it's got to be less than the above suggests.

Second, and as before, it's important to contextualize the wrong relative to all others. What we do to animals *is* terrible, to be sure. But so are many *other* things we do. It's *very bad* that so many people are still suffering and dying from preventable diseases; it's *very bad* that the contemporary West has, in effect, made indentured servants of developing countries; it's *very bad* that most of us can't be bothered to address systemic racial injustices in our own communities. Moreover, we've got better grounds for being

---

5. For the USDA's numbers (which don't include fish), see http://www.nass.usda.gov/. For the UN's global numbers (which include all species), see the *FAO Statistical Yearbook*, available here: http://www.fao.org/docrep/018/i3107e/i3107e03.pdf.

incensed by these injustices than we do the injustices faced by animals. Human societies have come a long way toward affirming human rights and the equal worth of human persons. Hence, it's especially appalling that we're so far from realizing these ideals. By contrast, most human societies are just beginning to give serious moral consideration to the animals we raise for food. (The rest aren't even doing that.) So, while our treatment of animals is gravely disappointing, it isn't at all surprising. Hence, the severity of the wrong is mitigated—at least with respect to what it's reasonable to demand— by the moral commitments of those to be confronted. The point here is not that eating an animal is less significant than other wrongs. Rather, it's that we wanted an argument for thinking that abstaining from meat-eating stands out among the various changes that people ought to make, that it's reasonable to ask for this change over others. I'm contending that this isn't obviously so.

Limiting the Context

The third way to argue that it's reasonable to demand that people change their diets is to limit the context. Perhaps it would be unreasonable to demand that strangers alter their behavior. Still, someone might argue, we can demand that our friends change; we can push our loved ones to do right by animals.

I think this is correct, but we shouldn't overstate its significance. Perhaps I can reasonably demand that my friend, the lapsed vegan, get back on the wagon. But it doesn't follow from this that I can demand the same from my brother, who has never shared the relevant ethic. It isn't enough to have *standing* to make demands, as we often do with respect to kith and kin; we have to have standing to make that *particular* demand.

At this juncture, the right move is to combine strategies. "Granted," someone might say, "it would be unreasonable to make demands of strangers. But we can reasonably demand that those near and dear become vegetarians, given both the ease of giving up meat and the severity of the wrong involved in consuming it."

I think this move fares no better. Consider a parallel line of reasoning: "Granted, it would be unreasonable to make demands of strangers. But we can reasonably demand that those near and dear give up dining at restaurants, redirecting that money to causes that desperately need it. After all, it's easy to give up dining out, and there are many evils we perpetuate by not giving." Again, the goal is to distinguish meat-eating from the other changes we might demand. And so far, we have not identified distinguishing features.

Defending the Argument's Second Premise, Step 3

If the arguments in the last two sections are sound, then we have yet to find a reason to demand abstaining from meat that doesn't apply equally well to a whole host of morally important changes. Since it would be unreasonable to demand them all and arbitrary to demand just one, this means we haven't yet found a reason to demand abstinence.

This raises an important question: what *would* it be reasonable to demand? I doubt we'll get a clean distinction between those norms where insisting on compliance would and wouldn't be reasonable. More realistically, we can hope to develop a very rough ranking of norms based on various considerations: those in the highest group will be the ones where it's clearly reasonable to express anger when people fail to comply with them; those in the lowest will be the ones where it clearly isn't reasonable. To this end, I propose five guidelines:

1. It's easier to justify demanding compliance when failure to comply with a norm somehow threatens our shared life. That is, when failure to comply undermines mutually beneficial cooperative activity, it's easier to justify expressing anger about the failure.
2. It's easier to justify demanding compliance when there's already widespread compliance. This is evidence that the compliance is within the person's capacity, that it isn't overly burdensome.
3. It's easier to justify demanding compliance when the violator endorses the norm in question. This allows us to borrow the violator's authority (at least insofar as the violator wouldn't want to resolve the inconsistency either by embracing it or by rejecting the norm that gave rise to it).
4. It's easier to justify demanding compliance when the norm in question is easy to apply. In part, this is because this feature makes it less plausible that the violation can be explained away as an accident.
5. It's easier to justify demanding compliance when we can be quite confident in identifying failures to comply, thereby lowering the risk of a false accusation.[6]

To see how we might deploy these considerations, let's think about theft. Is it reasonable to demand that people not steal? Absolutely. First,

---

6. These considerations are loosely based on the ones that J. O. Urmson proposes—for quite different purposes—in Urmson (1958, pp. 211ff.). Urmson's goal is to defend a distinction between duty and the supererogatory, and he argues for a conception of morality that serves "man as he is and as he can be expected to become, not man as he would be if he were perfectly rational or an incorporeal angel" (1958, p. 210). As it happens, I have little sympathy with that project. For challenges to Urmson, see Singer (1972), Pybus (1982), Kagan (1989), and Unger (1996).

non-compliance undermines various aspects of our shared life: in addition to the obvious material losses, it also undermines trust, forcing us to redirect energy from constructive projects to security measures. Second, compliance is indeed widespread (at least in societies where there are reasonable opportunities for securing the means of existence through licit channels), so we have reason to think that we aren't asking too much. Third, the norm in question is one that we can expect people to endorse; we all want to claim a right to control our property, and we all enjoy the goods that compliance makes available to us both individually and collectively. Fourth, the relevant rule—"Don't take what doesn't belong to you"—is not unmanageably complex. And fifth, we can often be quite confident about failures of compliance, at least when the perpetrator is in possession of the goods in question. In this context, demands are completely reasonable.

By contrast, consider the imperative to abandon your projects and devote your life to the needs of malnourished children in the developing world. Presumably, even if we think we ought to do this (as I often do), we don't think it would be reasonable to demand this of others. And the above considerations explain why. First, our failure to do this doesn't undermine some aspect of our shared life together (where, again, "we" are healthy Western people who live well above the poverty line). Second, there certainly isn't widespread compliance. Third, most people don't endorse the norm. Still, the imperative does quite well on the fourth and fifth considerations: while it isn't easy to specify what compliance involves, we do have a fairly good sense of what it would mean to be in the ballpark; and, if nothing else, it's pretty easy to identify what it *doesn't* involve. So, we can be quite sure that failure to comply isn't accidental, and that we aren't misattributing responsibility. However, it's plain that these virtues don't trump the other vices. Perhaps the most natural way to make sense of this is to understand the fourth and fifth considerations as defeaters, rather than positive considerations: that is, they can weaken an existing case for the reasonableness of a demand, but they aren't sufficient to make the case on their own.

Of course, none of this shows that stealing is worse than failing to devote your life to those in need, nor that a life of (relative) luxury is indeed morally permissible. Rather, these cases are designed to illustrate the difference between reasonable and unreasonable demands. Based on where we are socially and morally, insisting on compliance seems perfectly reasonable in the one case, and far too strong in the other.

Is failing to abstain from animal flesh more like theft, or is it more like failing to devote yourself to the distant needy? Well, it's hard to see how it undermines our shared life. There definitely isn't widespread compliance. Most don't endorse the norm against it. And while it fares reasonably well on the fourth and fifth guidelines—"Don't eat meat" is pretty straightforward as imperatives go, and it's not hard to tell when someone's having a steak—these considerations aren't enough on their own, at least if I'm right to think of them as defeaters. So, failing to abstain from animal flesh seems to be much like failing to devote yourself to the distant needy. It may be morally required, but it isn't reasonable to demand.

Where does this leave us? Recall the main argument:

1. If it would be unreasonable to demand that someone behave in a particular way, then, if she fails to behave that way, we shouldn't blame her for it.
2. It would be unreasonable to demand that someone abstain from eating meat.
3. So, if someone eats meat, we shouldn't blame her for it.

I've assumed the truth of the first premise, which I take to be plausible in its own right. In this section and the last two, I've argued for the second. If these arguments succeed, then the conclusion follows.

An Objection

At this juncture, someone might want to revisit the argument's first premise, criticizing it as follows. Surely there have been times in history when, for the sorts of reasons discussed above, it would have been unreasonable to demand that someone free his (human) slaves. Nevertheless, someone might insist, it would have been permissible—if not morally mandatory—to blame him for not doing so. Hence, reasonableness can't be the line between permissible and impermissible blame.

There are three things to say about this objection. The first is a concession in the direction of the objection: I grant that *some* resistance to slavery has always been morally permissible, and perhaps morally mandatory. However, it doesn't follow from this that the relevant form of resistance is blame. There are, of course, other means available to us: moral argumentation, pleading, humor, leading by example, non-violent protest, and so on. Unless we assume that we ought to blame people for every blameworthy action—a position I reject here—it isn't obvious to me how to make the move from resistance to blame.

The second is this. We are inclined, in hindsight, to see the actions of reformers as entirely justified. Perhaps this is evidence of a consequentialist bias in our thinking about the behavior of historical figures. Either way, though, I think it isn't right—or, at least, isn't clearly right. It may be the case that many successful methods of moral reform are themselves morally suspect. Violent revolution may sometimes secure social change, and we should be glad for the change. But it does not follow that the violent revolution was justified. (This could be because there were other, non-violent means available. Or it might be because Kant was right about there being categorical imperatives. Surely there are other possibilities too.) So, I'm inclined to bite the bullet here. Granted, blame is a far cry from violent revolution; still, I suspect that there *were* eras in which it would have been unreasonable, and therefore wrong, to demand that someone free his slaves—unless the demander is himself the slave. (We might chalk up the counterintuitive character of this position to our moral distance from such eras.) Nevertheless—and this is crucial— slavery has always and everywhere been deeply wrong, and wherever there have been slave owners, they should have freed those they'd enslaved.

Third, we should remember that for every moral battle we fight, we ignore others. We wrong both people and animals in all sorts of ways. (For example, we're guilty of far more than *eating* animals. We wrong them by abusing them before we kill them, by testing on them, by crushing them with our cars, by paving over their habitats, by wreaking havoc on the environment that sustains them.) With this in mind, we mustn't become myopic in our moral focus. I want moral progress in a thousand areas, and I'll take it where I can get it. In- sisting on one dimension of that progress is counterproductive in at least this way: it blinds us to other goods that we're in a better position to secure. I'm often surprised by how negatively people view vegetarians—as self-righteous, immature, effete, insufficiently concerned with human suffering. Moral argu- mentation tends to require moral authority, and it's a mistake to sacrifice it un- necessarily.[7] We are many, many years away from the day when our society will see meat-eating as calling for blame. Until then, I see no reason to lose the other battles—for human and nonhuman animals alike—that we might still win.

Conclusion

If the arguments against meat-eating are successful, then we can't justify sig- nificant costs to others with insignificant benefits to ourselves; but since so many of our decisions involve such a tradeoff, so many of them are morally

---

7. For that very reason, vegetarians shouldn't compromise their vegetarianism for the sake of public perception.

wrong. However, if much of what we do is morally wrong, then it would be unreasonable to demand that people make every moral change that's required of them, and it would be arbitrary to insist on any one—at least without explanation. This shifts the burden of proof; we now need an argument for demanding compliance with the "Don't eat meat" norm. However, the arguments available seem to apply just as well to a host of other norms. Moreover, when we take stock of the considerations relevant to when a moral demand is reasonable, we see that insisting on abstinence from meat-eating doesn't fare so well. But if it would be unreasonable to demand that someone behave in a particular way, then we shouldn't blame her for failing to behave in that way. So, if someone eats meat, we shouldn't blame her for it.

Before wrapping up, let's return to the passage at the beginning of this chapter, drawn from the original preface to Singer's *Animal Liberation*. When Singer says he hopes that, after reading his work, we will experience "emotions of anger and outrage," he doesn't say how he hopes we'll direct those emotions. Given the rhetoric he employs—comparing speciesism to racism and sexism and pointing to the parallels between our treatment of animals and human slavery—it's not unreasonable to think that he expects some of this anger to be directed toward those who perpetuate a system in which so many animals are anything but free. If this reading is correct, then the argument I've made puts me at odds with him.

It does seem to me that anger has its place. And if we keep reading, I think we find another way to read the passage that's compatible with the conclusion I've defended. Singer says that he hopes these strong emotions will be "coupled with a determination to do something about the practices described." Perhaps Singer is thinking of anger as a force that can motivate us, as something that compels us to address injustice. Anger need not be directed at persons to accomplish this, but at the world as we find it, full of ugly realities and the mechanisms that protect them. This frustration can drive us to be agents of change—not by blaming individuals, but by working to dismantle the structures that support mindless cruelty. (This is, perhaps, the difference between vegan street activism and legislative reform.) So I can, I think, affirm Singer's hope for anger and outrage, albeit not toward those who eat meat.[8]

---

8. Thanks to Craig Hanks for encouraging me to develop the idea behind this chapter. For helpful conversations and feedback on drafts, thanks to Ben Bramble, Brian Coffey, Jeff Gordon, Jeff Johnson, Robert Jones, Rima Kapitan, Howard Nye, and Burkay Ozturk.

References

Adams, Carol. 2010. *The Sexual Politics of Meat: A Feminist-Vegetarian Critical Theory, 20th Anniversary Edition*. New York: Bloomsbury Academic.

Barnard, Neal, and Kristine Kieswer. 2004. "Vegetarianism: The Healthy Alternative." In *Food for Thought: The Debate over Eating Meat*, ed. Stephen F. Sapontzis, 46–56. Amherst, NY: Prometheus Books.

Curnutt, Jordan. 1997. "A New Argument for Vegetarianism." *Journal of Social Philosophy* 28 (3): 153–172.

Degrazia, David. 2009. "Moral Vegetarianism from a Very Broad Basis." *Journal of Moral Philosophy* 6 (2): 143–165.

Engel, Mylan. 2001. "The Mere Considerability of Animals." *Acta Analytica* 16: 89–107.

Hursthouse, Rosalind. 2000. *Ethics, Humans, and Other Animals: An Introduction with Readings*. London: Routledge.

Hursthouse, Rosalind. 2011. "Virtue Ethics and the Treatment of Animals." In *The Oxford Handbook of Animal Ethics*, ed. Tom L. Beauchamp & R. G. Frey. New York, NY: Oxford University Press.

Jordan, Jeff. 2001. "Why Friends Shouldn't Let Friends Be Eaten: An Argument for Vegetarianism. *Social Theory and Practice* 27 (2): 309–322.

Kagan, Shelly. 1989. *The Limits of Morality*. New York, NY: Oxford University Press.

Kriegel, Uriah. 2013. "Animal Rights: A Non-Consequentialist Approach." In *Philosophical Perspectives on Animals*, ed. K. Petrus and M. Wild, 231–247. Bielefeld, Germany: Transcript.

Midgley, Mary. 1984. *Animals and Why They Matter*. Athens, GA: University of Georgia Press.

Nobis, Nathan. 2002. "Vegetarianism and Virtue." *Social Theory and Practice* 28 (1): 135–156.

Norcross, Alastair. 2004. "Puppies, Pigs, and People: Eating Meat and Marginal Cases." *Philosophical Perspectives* 18 (1): 229–245.

Pybus, Elizabeth M. 1982. "Saints and Heroes." *Philosophy* 57 (220): 193–199.

Rachels, James. 1977. "Vegetarianism and 'The Other Weight Problem,'" in *World Hunger and Moral Obligation*, ed. William Aiken and Hugh LaFollette, 180–193. Englewood Cliffs, NJ: Prentice-Hall.

Rachels, James. 2004. "The Basic Argument for Vegetarianism." In *Food for Thought: The Debate over Eating Meat*, ed. S.F. Sapontzis, 70–80. Amherst, NY: Prometheus Books.

Regan, Tom. 1983. *The Case for Animal Rights*. Berkeley, CA: University of California Press.

Singer, Peter. 1972. "Famine, Affluence, and Morality. *Philosophy and Public Affairs* 1 (3): 229–243.

Singer, Peter. 2009. *Animal Liberation: The Definitive Classic of the Animal Movement*. New York: Harper Perennial.

Unger, Peter K. 1996. *Living High and Letting Die: Our Illusion of Innocence.* New York: Oxford University Press.

Urmson, J. O. 1958. "Saints and Heroes." In *Essays in Moral Philosophy*, ed. A. I. Melden, 198–216. Seattle, WA: University of Washington Press.

Walker, R. 2007. "Animal Flourishing: What Virtue Requires of Human Animals." In *Working Virtue: Virtue Ethics and Contemporary Moral Problems*, ed. Rebecca L. Walker & P. J. Ivanhoe, 173–189. New York: Oxford University Press.

Watson, Gary. 2004. "Responsibility and the Limits of Evil: Variations on a Strawsonian Theme." In *Agency and Answerability: Selected Essays*, 219–259. New York: Oxford University Press.

12 BEETLES, BICYCLES, AND BREATH MINTS: HOW "OMNI" SHOULD OMNIVORES BE?

Alexandra Plakias

Introduction

It is a familiar truism that humans are, by evolutionary design if not by choice, omnivorous. But this is not literally true: we don't eat, nor are we expected to eat, *everything*. In part this is due to physiological constraints: with the exception, perhaps, of Guinness-record holders, humans lack the capacity to ingest and digest such items as glass, automobiles, or bicycle tires. But much of what is sold under the label "food" in contemporary society lacks nutritional value, and many potentially nutritive foods are ruled out as "food," facts which complicate attempts to explicate what counts as food and why. In fact, "food" is not merely a descriptive label but a kind of normative status. Disagreements over whether or not eating meat is permissible can therefore be understood as disagreements over what should count as food. I'll suggest that by understanding the dispute between vegetarians and meat-eaters this way, we can situate it within a larger debate over how to conceptualize food; I'll go on to examine different dimensions along which we might distinguish food from non-food. What I'll suggest is that, while we typically think that the label "food" is best applied or withheld depending on the *kind* of thing in question, we ought instead to consider the processes a thing is subjected to when deciding whether to call it food. Viewing the issue in these terms helps draw our attention to the degree of autonomy we enjoy with respect to food—and the moral responsibility that comes with that autonomy. It also demonstrates that the debate over meat-eating is a worthwhile one to have, even if it turns out to be irresolvable—its moral value does not hinge on anyone's changing their mind about the permissibility of eating animals. What's at stake in the debate between vegetarians and carnivores isn't just meat: it's food.

Moral Disagreements

Some philosophers believe all moral disputes are, at least in principle, resolvable. That's because if two people disagree morally, one of them must be making some kind of error, and even if that error is extremely difficult to detect, given enough time, information, and cognitive powers, the parties to the disagreement will eventually discover it and thereby come to agreement. For example, I may disagree with your claim that it's morally wrong to kick puppies for fun, because I mistakenly believe that puppies are actually robots, and therefore incapable of feeling pain. In this case, once my mistake is corrected, I will come to agree with you about the wrongness of kicking puppies. Virtually no one denies that some moral disagreements are like this—many moral disputes have been based on faulty factual beliefs, and when these factual beliefs are corrected, the dispute is resolved. What is controversial is whether *all* moral disputes are like this.[1]

I'll be interested in a different kind of dispute, one that turns not on questions of fact but on more difficult issues about which concepts to employ. I'll suggest that debates over meat-eating are like this; I'll also suggest that we view them not as aimed solely at resolution, but at realizing a distinctively moral kind of value. What might this mean? Defining moral value is difficult, and I won't attempt to defend a particular definition here. But a comparison with epistemic value might be useful. As a first pass, we might say that a dispute has moral value if it makes us better moral agents, if it helps us fulfill our moral obligations, or if it produces some moral good. I'll return to this issue later in the chapter, and discuss it with reference to the debate over eating meat. First, a few more words about the kinds and causes of moral disputes are in order.

---

1. For further discussion, see Doris & Plakias 2008.

As we saw above, some moral disputes turn on matters of fact; these can be resolved, even if only in principle, by bringing both parties into agreement on the relevant facts. Others may arise because some party or another is being irrational, or is being selfish, or is failing to appraise the situation properly. But some disputes may arise from a difference in concept. The dispute over the moral permissibility of abortion may be like this. What seems to divide the two sides is a difference in the concept of a person. One side believes that a fetus is a person, and therefore that abortion is the killing of a person. Another side believes that a fetus is a collection of cells—and that a collection of cells cannot be a person. Is this a factual difference? Certainly some disputes over abortion come down to factual claims about a fetus's ability to feel pain or its viability outside the womb. But at bottom, the question of what to count as a person is a conceptual question, not a factual one. We may use factual considerations in deciding the issue, but we will also have to use normative considerations. For example, should intelligent non-human primates count as people? Biology alone can tell us a lot about the intelligence of primates, but it can't answer that question for us; answering the question requires weighing the importance of various factors in determining personhood.

Below, I'll suggest that disputes over the permissibility of eating meat can be viewed as disputes over conceptual questions. But first, I need to address one potential line of objection. Some philosophers have argued that disputes that arise because two parties are using different concepts are really disputes about which terminology to use. Indeed, philosophers have referred to such disputes as "merely verbal"—note the derisive connotations—and have argued that they are not, therefore, worth engaging in (see, e.g., Sosa 2005). I think this view is mistaken. Sure, if we are arguing over whether Paul McCartney is a star, and you are using "star" to mean "a celestial body," and I am using it to mean, "a famous person," this is a pretty silly dispute to be having. As soon as each of us realizes what the other means, we can happily resolve our dispute—both are perfectly respectable uses of "star," and only one correctly applies to Paul McCartney. So here the dispute disappears when we realize that it is a dispute caused by confusion about words. But not all verbal disputes will be "merely" so—some may be worth having. Perhaps we are disputing whether someone is a good philosopher, and we cannot seem to agree. It emerges that your standard of "good," for a philosopher, is to be equal in fame and stature to Socrates—nothing else will do. Here, we are using "good philosopher" to mean different things, so the dispute is, arguably, a verbal one. But I do not foresee a happy resolution. After all, your concept is ridiculous! On your view, there have only been, perhaps, two or three good philosophers in all of history! And not only do I find your concept laughable, it may even be practically pernicious—if, for example, we are trying to hire someone to fill a key position, and you reject all the candidates because none of them are "good philosophers." The dispute is verbal, but that doesn't mean both parties have equal claims to truth here. Your concept is simply a *bad concept* to be using, and it seems well worth my while to dig in my heels and insist that I am in the right here (for more discussion of this point, see Sundell 2011). Thus, contrary to what some philosophers have claimed, it is not the case that the realization that a dispute is due to terminological or conceptual differences gives us reason to abandon it—especially not when practical matters hang in the balance. We can have meaningful and useful debates over which criteria ought to guide our application of a certain term or concept. And as we shall see in the next section, this is exactly what we find when we turn to the debate over the moral permissibility of eating meat.

Food Fights

We think of the debate over eating meat as a debate over what foods to eat. But what if, instead, we reconceive it as a debate over what food is and is not? That is, rather than think of the relevant question as "What foods are ethically permissible to eat?" we could think of the question as "What is and is not a food?" Reframing the issue in these terms situates it within the larger debate over what we ought to eat—and what criteria we ought to use to rule out certain items from the domain of food. I'll discuss two such criteria and suggest that they correspond to two different motivations for vegetarianism. Once we understand the debate as concerned with our conception of food, we can see how it offers the opportunity for moral progress even if there is, ultimately, no consensus on whether or not to eat meat. We can also become more attentive to our role in shaping the food choices available to us and others and the moral consequences of those choices.

One motivation for this reconceptualization of the debate is the recognition that the label "food" is both normatively significant and a function of our attitudes and choices. By labeling something "food" we implicitly assent to the idea that it's something to be eaten. This might seem odd—isn't food a descriptive label? When I say that something is food, aren't I merely describing it as having certain properties, such as being nutritious or digestible? In fact, a look at how we use the label "food," and the things we apply it to, reveals that this isn't the case. While scientists define food as substances that contain nutrients essential for life and are consumed to sustain life, this definition seems ill-equipped to help us navigate the eating habits of the developed world.[2] For one thing, much of what we consume to sustain ourselves has little to no nutritional value (diet soda, sugar-free candies) or represents unnecessary caloric intake (how many of us have mindlessly munched on some chips or popcorn while watching television?). And our intention when consuming food is rarely sustenance of life—we eat as a form of socializing, we eat to experience new or familiar things, we eat to be polite, we eat from boredom, or from curiosity . . . as the saying goes, in many developed societies today, we live to eat, rather than the other way around. So the scientist's definition, while perhaps helpful to those interested in studying the feeding behaviors of various species, is inadequate for those who are interested in studying what *people* eat, and why.

Even the agencies charged with regulating it have trouble defining what food is, except in terms of what we take food to be. The United States Food and Drug administration (2010) defines food as "articles used for food and drink" and also chewing gum. The European Union defines food as "any substance or product, whether processed, partially processed, or unprocessed, intended to be, or reasonably expected to be ingested by humans" and also includes chewing gum in this category (European Parliament, 2002). Notice that unlike the scientific definition we looked at earlier, these definitions make no reference to the nutritional value of food or to its contribution to sustaining life. (The inclusion of chewing gum makes it pretty clear that these are not requirements that something actually be food!)

What these definitions *do* share is this: they all define food in terms of what we consume or ingest. That is, part of what makes something food, rather than just a plant or an animal, is that we regard it and treat it as such. We don't eat corn because it's food; it's food because we eat it. Of course, the fact that corn thereby becomes food makes us much more likely to eat it— which in turn further entrenches its status as food, thereby making it all the more likely to be eaten . . . and so on. The process by which an edible substance becomes food is, well, eating.

---

2. Most of what I'll say in what follows is directed at industrialized societies, where our problem is overconsumption of food rather than starvation. Of course, within these societies, the extent to which people have choices about what to consume varies a great deal, so the statements I make will not apply to all Americans, or all Western societies, or all industrialized societies. Whether, and to what extent, ethical obligations regarding food depend on one's material resources is an interesting and important question, but I can't take it up here.

Of course, not every substance that is eaten becomes food as a result. Michel Lotito, a Frenchman aptly nicknamed "Monsieur Mangetout," has ingested everything from airplanes to robots to bicycles (broken down into small parts, of course). Individuals with a condition known as "pica" experience urges to consume clay and dirt (among other substances). And of course, in certain contexts, humans consume one another's bodily fluids (saliva, semen). But in none of these cases does consumption render the consumed object or substance *food*. A lover may be happy to exchange saliva in some contexts, but spit on their morning muffin, and they are going to be less than overjoyed. This further reinforces the point that "food" is not merely a descriptive label: what's required for something to become food is not just a behavior—consuming it—but also a certain attitude toward it. (This also explains why items consumed under extreme duress, e.g., in times of extreme hunger, don't therefore become food. Seventeenth-century European writers describe instances of autophagy in times of extreme hunger, but they do so in a very different tone than that used to describe willing cannibalism. See Camporesi 1989.)

This feature also reveals the normative significance of labeling something "food." To say that something is food is to say more than that it is edible. There are many substances in our environment that would be relatively harmless to ingest, and are perhaps quite tasty, but which we would never think of eating; conversely, many of the things that we do eat are actually pretty bad for us. To say that something is food is, rather, to grant that we are prima facie justified in eating it; its status as food gives us a kind of permission. The act of eating an orange needn't be justified or explained; the act of eating orange clay does. Even if both are occasioned by a craving, the former is treated as normal, the latter as pathological. A similar distinction applies to reasons for rejecting foods. If, at a restaurant, I find a grub in my salad, I don't need to justify my decision not to eat it—it's not food (at least, not to us Westerners—but we'll get to insects in a bit). But if I refuse the salad, I may be asked for a reason. The fact that something is not a food means that the question, "why won't you eat that?" represents a kind of category mistake—it's just not the kind of thing one eats! On the other hand, the fact that something is a food licenses certain kinds of reasons for rejection. Typically, these involve considerations of taste (I hate hard-boiled eggs) or of health (I'm allergic to strawberries; I'm lactose-intolerant; I'm watching my cholesterol). These sorts of reasons are personal, insofar as we don't expect others to share them, and we don't try to impose them widely (though we may ask a host to refrain from cooking with nuts, or encourage someone to try a food they claim not to like, or ask someone not to eat hard-boiled eggs in front of us). My dislike of eggs is not a reason for you to reject them.

Why "Food" Matters

I've suggested that whether or not something is food not only affects how likely we are to eat it, but influences the kinds of reasons we think are (and aren't) required for eating (or not eating) it. To say that something is food is to say that it is the kind of thing that is acceptable to eat; it's to sanction eating it. And given that what and how we eat has morally significant consequences, calling something food also has morally significant consequences. The essays contained in this book demonstrate the moral implications of one of those choices: meat-eating. Even for vegetarians, the choice of what sorts of foods to consume has moral significance: conventional farming techniques have environmental impacts, as does the transportation of produce across long distances; both organic and non-organic produce is often picked by laborers working in conditions that amount to slavery; the use of genetically modified ingredients by food manufacturers—particularly those in soy products— influences what sorts of seeds farmers plant and how they must alter their farming practices. In 2001, Monsanto pulled its genetically modified New-Leaf potato seeds off the market, after McDonald's, responding to perceived consumer opinion, said it wouldn't sell GMO potatoes. And of course our food choices affect our health, which in turn has personal as well as social costs: as our concept of food has broadened, we've consumed more and more highly processed foods like hot dogs and Pop-Tarts—after all, they're food!

This last point illustrates one of the potential dangers of the label "food": it can lead us to feel justified in making choices that are harmful both to ourselves and to society. When highly processed meat and dairy products are sold alongside their (relatively) unprocessed counterparts, we are encouraged to think of them as the same kind of thing, and therefore to adopt the same attitude toward them—you might prefer one thing to another, but that's a personal preference. At root, they're both foods, right? And yet when the journalist Michael Pollan, a prominent critic of the contemporary food system, urges his readers to "eat more food," he is not encouraging them to go for another Big Mac or Twinkie (2007).

But it's not just the consequences of its application that make "food" a morally significant label. To call something—especially a living thing— "food" is to assign it a certain moral status. In some cases this assignment may appear unduly degrading. In the United States and most of Europe, for example, we think of dogs as pets. We coddle them, cuddle them, feed them—but never eat them. To call a pet dog "food" would be to somehow reduce its moral standing to a sort of object; to *eat* a pet dog, even if it died in an accident, is something most of us would find morally objectionable (see Haidt 2001 for empirical evidence to this effect). For a more dramatic example, consider the moral outrage that often accompanies reports of (non-starvation-induced) cannibalism. What is it about this practice that seems so objectionable to us? I submit that we object to the treatment of human flesh as mere *food*.

In other cases, we may object to labeling something food on the grounds that it would elevate something we consider base or unworthy of being eaten. Take, for example, insects. Insects are nutritionally superior to animal flesh in many respects. From an environmental perspective, they are vastly superior: they require far fewer resources to farm, and they emit far fewer waste products. Furthermore, because they're cold-blooded, insects can be frozen before they're killed, which puts them in a sleep-like state before death; they can therefore be killed with a minimum of suffering. And because they require less room, it's possible to raise many insects without causing suffering due to confinement or overcrowding. Thus rearing insects for food avoids many of the moral issues surrounding raising animals for food.[3] Indeed, the United Nations Food and Agriculture Organization recently released a report urging the adoption of entomophagy (eating insects); as the report bluntly states, "what we eat and how we produce it needs to be re-evaluated . . . we need to find new ways of growing food" (2013: ix). There are persuasive moral and environmental considerations in favor of eating insects: we can feed many more people and reduce both pollution and the consumption of valuable resources (in the form of grain and water, as well as the cost of raising and transporting animals) by switching to insect-based protein rather than animal-based protein. There may even be aesthetic reasons to consume insects: in many cultures they are considered a delicacy. Aristotle himself, in his *Historia Animalium*, favorably comments on the sweet taste of cicadas; the females, he notes, are sweetest after "copulation," when they are full of eggs.

What's stopping us? If asked, most of us Westerners would probably evince a kind of disgust at the thought of eating insects on the grounds that they're not *food*. (I confess to feeling a bit squeamish at the thought myself.) That is, we reject the thought of eating insects because of the *kind* of thing they are: dirty, disgusting, polluting. But can this reaction be justified as

---

3. For a moral argument in favor of entomophagy, see also Meyers 2013.

anything more than a prejudice? Consider: many of the features that make insects so unappealing—exoskeletons, antennae, their segmented bodies— are present in some of the most expensive and prized foods in Western cuisine.

Certainly many of the insects we encounter in our daily lives—especially if we are urban dwellers—do fit this description and are unfit to be eaten. Even the most enthusiastic proponent of entomophagy is not arguing that we trap and eat cockroaches from New York City streets! But this is a consequence of the environment in which these insects grow and are found. Those who advocate eating insects are interested primarily in how we might breed insects for this purpose—Mansour Ourasanah, in collaboration with KitchenAid, has recently designed the "Lepsis," a home system for raising grasshoppers for food.

We can't change the kind of thing insects are—we could disguise them as crackers or as candy, but a bug is a bug. What is interesting about the approach taken by Ourasanah is that it doesn't attempt to deflect attention from the fact that what's being eaten is a grasshopper, but instead draws our attention to the manner in which that grasshopper—that food, even—is raised and produced. Our refusal to consider insects the proper *kind* of things to eat may, in the end, stem from a kind of confusion between disgust at the origin or environment of the thing and the thing itself. And the result of this confusion is that we overlook or rule out a food source that is morally, environmentally, economically, and perhaps even aesthetically superior to much of what we choose to consume.

In sum: what we call "food" matters. It has important personal, social, and moral consequences. And with every choice we make about what to eat or not eat we incur some responsibility for those consequences. Unfortunately, our active role in defining the label "food" and deciding what it gets applied to is, all too often, opaque to us.

Why Meat Matters

So far I've been discussing food generally, rather than meat specifically. In the remainder of the chapter, I'll turn my attention back to the debate over meat-eating. What I will suggest is that the debate over whether to eat meat be understood as a debate over how to characterize food: specifically, whether meat ought to be considered food, and for what reasons. I'll then go on to argue that regardless of whether the debate can be resolved (I doubt that it can, for reasons that should become clear), it is nonetheless a morally valuable one to engage in—in fact, it's one that both meat-eaters and vegetarians alike have a moral responsibility to engage in.

First, a clarification is in order. When I talk about vegetarians, I'll be talking about those who refrain from eating meat for moral reasons. Some people may be vegetarians for health reasons, or for practical reasons; that's not the kind of vegetarianism I'm interested in here. I'm also not talking about the kind of vegetarian who simply dislikes meat, as a personal preference.

We can distinguish between two versions of (moral) vegetarianism. The first is what we might call "extrinsic" vegetarianism. On this view, there's nothing *intrinsically* wrong with eating meat; rather, it's the way meat is raised, slaughtered, and processed that makes it morally impermissible to eat. Peter Singer, for example, objects to meat-eating on the grounds that it causes unacceptable suffering in animals; presumably, if no animals suffered as a result—if we ate only animals that died of natural causes—Singer would not object.[4] The second kind of vegetarianism, which we might call "intrinsic" vegetarianism, holds that there's something intrinsically wrong with eating meat; under no conditions would it be morally permissible to eat meat. They don't just object to the process by which meat is produced, they object to the very idea of eating meat.[5] For intrinsic vegetarians, what's wrong with the idea of meat is that it involves eating the wrong *kind* of thing—sentient beings. Indeed, such vegetarians might well object to the term "meat" itself, connoting as it does not a living creature, but a food product. And that's the real issue between intrinsic vegetarians and meat-eaters: for the former, animals are simply not the right sorts of thing to be considered food.

Even the most committed carnivores typically draw the line at some types of meat. We don't see a peacock and think "dinner," as sixteenth-century Europeans might have done. For some of us, the line is drawn at horses (recall the horror with which the UK reacted to recent news that their frozen hamburgers and lasagna may have contained up to 30% horsemeat); for others, dogs; for all but a very few of us, humans. (Though the media has recently seized on the "trend" of mothers consuming their newborns' placentas; whether or not this counts as meat, I leave as an exercise for the reader.) What the intrinsic vegetarian insists is that we draw this line differently:

---

4. See, for example, Singer 1975; for another example of what I've called "extrinsic vegetarianism," see also Rachels 1977.

5. See, for example, Regan 1983; for an interesting discussion of related issues—though not a defense of vegetarianism per se—see Diamond 1978.

around all animals, perhaps, or all sentient creatures.[6] Thus, the dispute between intrinsic vegetarians and meat-eaters is a dispute over whether sentient beings are food.

The carnivore who has trouble understanding the intrinsic vegetarian's rejection of cows and pigs as food would do well to consider their own attitude toward consuming dogs or guinea pigs or even humans. The intrinsic vegetarian is not stating a personal preference against eating meat and asking others to share it. They are not asking us to avoid eating animal flesh because of extrinsic facts about its production. Rather, they are asking us to change our mind about what food is—and what animals are. This is, at root, a dispute that turns on a conceptual difference.

What about the extrinsic vegetarians I mentioned above—those who object to meat-eating because of facts about the way that meat is farmed and slaughtered? In this case, it might seem that the dispute between extrinsic vegetarians and meat-eaters is *not* a conceptual one, since these vegetarians do not deny that it is permissible to eat meat in some circumstances— circumstances where the meat is obtained without causing pain or suffering to the animal—they just deny that we are actually in those circumstances. So for extrinsic vegetarians, the problem with meat isn't the meat but the means. The dispute isn't over what counts as food but over what counts as a morally permissible means of obtaining or producing it.

But why can't the production process play a part in whether or not something is food? I'm not talking here about certain processes necessary to render a substance edible or non-toxic, as when barley must be hulled in order to be edible, or cassava must be soaked and cooked to remove toxins. Rather, I'm talking about instances in which the same substance is either regarded as food or not, depending on how it has been treated or produced. In fact, some of our judgments about whether or not something is food are already guided by how that thing is produced or treated, whether or not we realize it. We eat a deer whose meat is bought at the farmer's market, but if we come across that same deer on the side of the road, after it's been struck by a car, it's roadkill— not food.[7]

If we can rule out the possibility of eating something on the grounds of, that's not what food is, why can't we rule out eating something on the grounds

---

6. Some vegetarians think it's acceptable to eat certain kinds of mollusk, like oysters and mussels. Others refuse any animal at all.

7. See Bruckner (this volume) for an argument to the effect that we are, in fact, morally *obligated* to eat roadkill.

of, that's not where food comes from? Most of the animals sold and consumed as food in the developed world have been bred to maximize the amount of meat—even at the expense of their ability to move and mate—and have been given hormones, antibiotics, and other drugs, partly to ameliorate the effects of the overcrowded and inhumane conditions they're raised in. Is denying that the resultant meat is food really that different from denying that non-dairy creamer, or a sugar-free lifesaver, is food?

In fact, many carnivores do make discriminations like these. As Americans have become less and less inclined to view unprocessed offal as food, produc-ers have looked for ways to transform those parts of the animal into palatable creations such as hot dogs, bologna, McRib sandwiches, and so on.[8] And yet when we're confronted with the process behind such products, we repudiate them: witness the recent outrage at the "revelation" (though it should hardly have been surprising) that many fast-food hamburgers were composed of so-called "pink slime": a sort of paste consisting of animal trimmings (official name: "lean finely textured beef trimmings") that had been treated with am-monia to kill salmonella and e. coli bacteria. Carnivores' outrage and disgust at this discovery might seem like hypocrisy or worse—if we're going to eat animals, we should eat the whole animal, rather than eating some "choice cuts" and letting the rest go to waste. But it might also be viewed as a kind of discrimination worthy of cultivation, insofar as it indicates that no matter what kinds of things we are willing to eat, there are certain kinds of processes whose result we cannot admit is food.

And here I think we can see room for, if not moral consensus, moral prog-ress. The intrinsic vegetarian argues that it is the *kind of thing* meat is, the fact that it comes from a sentient being, that makes it unfit to be considered food (indeed, they may even object to the label "meat," connoting as it does some-thing to be eaten). The extrinsic vegetarian argues that it's the *process* by which meat is obtained that makes it unfit. The carnivore thinks at least some animals are food, while perhaps denying that all parts of an animal are food. This gap may be unbridgeable. We may never come to consensus on what kinds of things count as food—nor, perhaps, should we want to. The diver-sity of cuisines across cultures and across history is, arguably, a good in itself; the United Nations Educational, Scientific, and Cultural Organization (UNESCO) has declared certain traditional cuisines (Mexican, Japanese, French) as part of our "intangible cultural heritage" and thus deserving of

---

8. Up until the eighteenth century, the French viewed the udder as one of the preferred parts of the cow; beef itself was slower to catch on. See Flandrin 2000, ch. 31.

protection.[9] But absent a consensus on what food is, we can make progress by attending to the processes by which we allow that food can be arrived at. In the final section I say more about the form such progress might take.

Food Technology, Food Transparency

Our attitudes toward food, as evinced by popular culture, display a striking ambivalence. On the one hand, the last decade or so has witnessed an increased desire for unprocessed, natural, and of course, whole foods. Sales of organic foods are through the roof, and even cereals like Froot Loops brag that they contain "whole grains." Domino's brags about their "handcrafted" pizzas; the McDonald's website invites visitors to "meet their suppliers" and brags that their milk is "from farm to restaurant" (neglecting to mention that perhaps there are a few steps in between). On the other hand, we live in an age of unparalleled technological innovation—20,000 new "food products" were introduced in 2010, according to the USDA. We have yogurt—sorry, "Go-Gurt"—that comes in tubes and candy-corn-flavored Oreos. The same company that brags about Froot Loops' "whole grains" markets a *breakfast* cereal called "Krave S'Mores" with a marshmallow and chocolate filling. We even flock to restaurants with "factory" in the name—not just The Cheesecake Factory, but also The Burger Factory, even The Food Factory. This ambivalence reveals a conflicted attitude toward the means by which our food is produced; we can no longer ignore the role that technology plays in our food supply, nor should we, as it has implications for the way we approach debates over the nature of food.

The role we accord technology also has implications for how we approach the debate over meat-eating. According to intrinsic vegetarianism, technology might well be the solution to the problem of meat-eating, since it can replace animal flesh with other substances, such as Quorn, tofu dogs, imitation lunch meats, and so on. Technology also promises new and improved versions of these products—Silicon Valley is investing tens of millions of dollars in developing new vegan "eggs" as I write this. Not to mention the possibility that we might someday have access to affordable, laboratory-grown meat (at the moment, a lab-grown hamburger costs about $100,000; at that price, it's unlikely to be making an appearance in the grocery store any time soon). These innovative substitutions for meat may represent the vegetarian's best

---

9. http://www.unesco.org/culture/ich/index.php?lg=en&pg=00002.

hope of eliminating animal flesh-based foods from our diet. On the other hand, insofar as they succeed because of their resemblance to meat, they may not represent a change in our attitudes about what food is so much as a willing suspension of disbelief.

On the other hand, if the objection to meat is based on the highly mechanized, industrialized process by which it gets to our tables (or cars, or desks—in 2012, about half the money spent on food was spent on food eaten away from home), then technological innovation may not be the solution, but instead part of the problem. If our conception of food is based in part on the process by which that food was made, then the ability to judge whether or not something is food requires a kind of transparency that highly processed foods may be unable to provide. As it currently stands, consumers often don't know where the meat they eat comes from or even how many cows the meat in a single hamburger comes from. The enzyme transglutaminase can be used to bind scraps of meat together into something that is visually indistinguishable from a steak.[10] If we reserve the label "food" for meat produced by non–factory farms, then consumers will have to seek out suppliers carefully, and only deal with butchers who are transparent about where they obtain their animals. A process- or production-based conception of food demands less distance between the source and the consumer; replacing meat with technological advances in food places more distance. And distance can be morally dangerous, insofar as it allows us to maintain moral blind spots.

Drawing a distinction between food and non-food with reference to process rather than kind is conducive to moral progress on several fronts. It allows us to overcome certain biases we have against foods that may be morally preferable to our current choices, such as roadkill and insects. It forces us to attend to the ways that what we eat is produced, which may seem obvious, but is something that consumers in the industrialized world rarely do. In doing so it opens up room for discrimination between apparently similar options: no longer are we choosing beef or chicken: we're choosing between beef and a factory-produced item. As consumers demand more transparency, corporations are likely to find that the most objectionable aspects of meat and animal product production are no longer economically viable.

---

10. Fortunately, the USDA requires that, when sold in stores, such products must carry a label informing consumers that they are buying "reformed" meat; unfortunately, there's no such requirement for restaurants, which, perhaps not coincidentally, is where most of these products are sold and consumed.

And, importantly, it makes us aware that we are active rather than passive agents in the construction of food: it reveals that we don't just make food; we make *food*.

Conclusion

I've presented two ways of drawing a line between food and non-food—in terms of the kind of thing it is and in terms of the process by which it's produced—as if they were opposed to one another, but they needn't be. In fact, even once one takes a stand on what kind of things are apt to be considered food, the question of how process and/or production should affect our choices remains wide open. My point here is not, primarily, that one way of drawing the distinction is in itself better than the other—though I have suggested that—but that we can characterize food both by what *kind* of thing it is, and by *how* it is produced or processed. Because of the way food is manufactured, regulated, and sold, this latter option is often invisible to us—we are encouraged to view cheese as cheese, whether in a can or in a block of cheddar. We're encouraged to choose food based on its nutritional content, but to pay no attention to the difference between the vitamin C in an orange or the vitamin C in Kool-Aid. The dimensions of kind and process, then, are not just grounds on which to reconsider the status of meat, but grounds on which to reexamine our treatment of food generally—and offer a way of navigating the ethical challenges posed by a world that presents us with tens of thousands of new "food products" a year.

References

Camporesi, P. 1989. *Bread of Dreams: Food and Fantasy in Early Modern Europe*. Chicago: University of Chicago Press.

Diamond, C. 1978. Eating Meat and Eating People. *Philosophy* 53: 465–479.

Doris, J., & Plakias, A. 2008. How to Argue about Disagreement. *Moral Psychology: Volume II*, Walter Sinnott-Armstrong, ed., 303–331. Cambridge, MA: MIT Press.

Flandrin, J. L. 2000. The Early Modern Period. In *Food: A Culinary History From Antiquity to Present*, Flandrin & Montinari, eds., 349–373. New York: Columbia University Press.

Haidt, J. 2001. The Emotional Dog and Its Rational Tail: A Social Intuitionist Approach to Moral Judgment. *Psychological Review* 108: 814–834.

Meyers, C.D. 2013. Why it is Morally Good to Eat (Certain Kinds of) Meat: The Case for Entomophagy. *Southwest Philosophy Review* 29: 119–126.

Pollan, M. 2007. Unhappy Meals. *New York Times Magazine*, Jan 28.

Rachels, J. 1977. Vegetarianism and 'The Other Weight Problem.' In *World Hunger and Moral Obligation*, Aiken & LaFollette, eds., 180–193. Englewood Cliffs, NJ: Prentice-Hall.

Sosa, E. 2005. A Defense of the Use of Intuitions in Philosophy. In *Stich and His Critics*, Bishop & Murphy, eds., 101–112. Oxford: Blackwell.

Sundell, T. 2011. Disagreements about Taste. *Philosophical Studies* 155: 267–288.

US Food and Drug Administration. 2010. *Federal Food, Drug and Cosmetic Act*. URL=http://www.fda.gov/RegulatoryInformation/Legislation/FederalFood DrugandCosmeticActFDCAct/default.htm.

INDEX

31901056997572